ISSN 1020-1203

Water Reports 25

AF863611

GROUNDWATER MANAGEMENT

THE SEARCH FOR PRACTICAL APPROACHES

A joint publication of the
FAO Land and Water Development Division,
the International Atomic Energy Agency,
the United Nations Department of Economic and Social Affairs
and the
United Nations Educational, Scientific and Cultural Organization

with contributions from the
Institute for Social and Environmental Transition
Boulder, Colorado
United States of America
and
the International Association of Hydrogeologists
Kenilworth, United Kingdom

FOOD AND AGRICULTURE ORGANIZATION OF THE UNITED NATIONS
Rome, 2003

ISBN 92-5-104908-4

Preface

This report grew out of a request to within the UN Administrative Coordinating Committee (ACC) Sub-Committee on Water Resources to establish a position on groundwater management within the group of UN agencies involved with groundwater development and management. It comes at a time when the use of the world's groundwater resources and related aquifer 'services' is intensifying, with little or no prospect for resolving the detrimental impacts through conventional management approaches. This raises the question as to whether such conventional water management approaches are inherently limited with respect to groundwater or whether the policy 'space' needs to be radically expanded to accommodate the drivers of demand for limited groundwater resources. The more specific question is then how the UN agencies involved with groundwater then position themselves to best effect in their respective country and regional programmes. Contributions were made from the ACC Sub-Committee members and the International Association of Hydrogeologists but a significant contribution to the report came from Marcus Moench at the Intitute for Social and Economic Transition (ISET) to whom the UN agency contributing authors are grateful. The relationship of the economic and social dynamic to the specific aquifer hydrodynamic is key. The rates of change of urbanization and agricultural intensification are now unprecedented. Complex patterns of production and consumption occur atop a set of vulnerable aquifers whose roles as physical and economic buffers are largely unrecognized. This report makes a plea for groundwater resource management to become much more realistic in what can be achieved in practice. This, it is argued, will need to be based on a much clearer appreciation of the social and economic drivers of groundwater use.

Contents

List of boxes

Acronyms

IAEA	International Atomic Energy Agency
IAH	International Association of Hydrogeologists
IFAD	International Fund for Agricultural Development
IHP	International Hydrological Programme
ISET	Institute for Social and Economic Transition
MWR	Ministry of Water Resources
NSAS	Nubian Sandstone Aquifer System
SCARP	Salinity Control and Reclamation Project
UNACC	United Nations Administrative Coordinating Committee
UNDESA	United Nations Department of Economic and Social Affairs
UNESCO	United Nations Educational, Scientific, and Cultural Organization
WWAP	World Water Assessment Programme

Chapter 1
Introduction

Warnings of a groundwater crisis (with falling groundwater tables and polluted aquifers) have led to calls for urgent management responses. However, much of the discussion has been on the basis of anecdotal evidence (Brown and Halweil, 1998; Postel, 1999; Sampal, 2000). There is a need to evaluate the hard evidence of there being such a crisis and to identify the types of management responses that actually work. Moreover, it is necessary to determine how the UN system should continue to respond. While not denying the severity and urgency of the problems facing groundwater management, there does arise the question as to how broad macro generalizations translate into specific management responses. Competent UN agencies maintain that these issues have to be addressed within the specific contexts of the hydrogeological settings (and particularly the hydrodynamics of the aquifer systems) and the patterns of human use to which they are subjected. Failure to recognize the variability and range of these physical limits (and the range of services that groundwater and aquifers provide), together with the breadth of social demands placed upon aquifer systems, will continue to result in ineffective management responses. In this sense, groundwater management is required to be highly localized, and to a far greater degree than that applied to surface water management (the effective management of perennial surface water flows usually occurs at greater spatial scales than the level of an individual borehole catchment).

Groundwater continues to serve as a reliable source of water for a variety of purposes, including industrial and domestic uses and irrigation. The use of generally high-quality groundwater for irrigation (which is largely indifferent to water quality) dwarfs all other uses (Burke, 2002). In many developing country settings, reliance has turned to dependency and the establishment of perceptions of access and use that are intensely 'private' irrespective of the legal status of the groundwater. However, groundwater and the aquifers that host it are inherently vulnerable to a wide range of human impacts. The development of mechanized pumping technologies in the mid-twentieth century has induced widespread drawdown externalities, including the depletion of the all-important shallow aquifers. The disposal of human and industrial waste and the percolation of pesticides and herbicides have degraded many aquifers beyond economic remediation. The largely unseen nature of groundwater has resulted in development initiatives that are unaware of the hydrodynamic limits of the resource and unable to regulate the resulting patterns of abstraction. The consequences range from the drawdown of water levels beyond the limits of dugwells and manual pumping technologies to more subtle and deferred environmental health impacts resulting from the migration of poor-quality water, e.g. the mobilization of naturally occurring arsenic by drilling deep tubewells in Bangladesh is well documented (http://www.bgs.ac.uk/arsenic/bangladesh/reports.htm).

FAO, the International Atomic Energy Agency (IAEA), the United Nations Department of Economic and Social Affairs (UNDESA), and the United Nations Educational, Scientific, and Cultural Organization (UNESCO) have identified groundwater management as a key theme in their programme of normative and operational activities. Over some 40 years of technical cooperation, FAO and the UNDESA have been involved intimately with groundwater development and management in developing countries. The IAEA is actively engaged in a programme of groundwater isotope analysis to determine the ages and provenance of groundwater in key aquifers throughout the world. Through its International Hydrological Programme (IHP), UNESCO has been the principal supporter of groundwater science and applied research, providing a bridge between advances in the developed world and demands from developing countries.

The purpose of this paper is to:

- establish links between the social and technical aspects of groundwater management against a contemporary background of rapid groundwater depletion and aquifer degradation;
- search for guiding principles and criteria for establishing more sustainable paths to groundwater management through practical actions;
- indicate a response of the UN system.

The paper uses case studies to illustrate key issues where source material is presented. It also proposes a research agenda to plug gaps in groundwater management. Furthermore, it elaborates strategic themes, principles and criteria in promoting sustainable groundwater management.

A contemporary search for principles needs to be based on the past 70 years of experience in both small and large-scale groundwater development (since the advent and popular deployment of mechanized borehole pumps). It should specifically address the areas of stress where congruent socio-economic and environmental factors are threatening long-term economic development.

With these issues in mind, the then UN ACC Sub-Committee on Water Resources (now simply called UN-Water) requested FAO, IAEA, UNESCO and UNDESA to prepare a paper on groundwater management for presentation to the Committee. Rather than providing technical solutions, the aim was to determine principles for policy responses and institutional adaptation in order to enable individuals, communities and economic agents to engage with groundwater in a fashion that could begin to close the gaps.

One of the first attempts to outline such principles led to the report titled "Large Scale Groundwater Development" (United Nations, 1960). The report was a response to the rapidly growing importance and exploitation of groundwater in a period of post-war reconstruction and expansion of irrigated agriculture. However, the world has moved on since then with many of the social and imperatives outweighing a considered, balanced approach to groundwater development and management. More recent approaches (World Bank, 1998, 1999) have developed these principles. However, such technical approaches start from the assumption that it is the resource that is to be managed rather than the use. This paper argues that it is also necessary to understand the 'pull factors': the demand for groundwater and aquifer supplied services to agriculture, municipal water supply, waste disposal and geotechnical stability (the problems associated with subsidence and groundwater rise). Furthermore, the paper contends that it is not possible to resolve competing demands for groundwater and aquifer use by simply managing the water resource. It argues that it is necessary to engage a broader array of socio-economic levers and actors.

Chapter 2
Groundwater degradation and the limits of hydrological information

The hard evidence required to assess global trends in groundwater depletion and aquifer degradation does not exist. A recent groundwater and food security study by FAO (2003) confirms that it is not possible to assess the extent to which global food production could be at risk from overabstraction. Indeed, the search for reliable groundwater-level and abstraction data (to determine depletion rates) was fraught with problems of coverage, consistency and reliability. The study concluded that it was not possible to obtain reliable time-series data on groundwater levels in specific aquifers in India and China in order to confirm or refute assertions about the threats to food production posed by groundwater depletion.

As a report by the Ministry of Water Resources (MWR) on a water strategy for the 3-H plain in northern China notes: "*Effective management (of groundwater) is highly dependent on appropriate reliable and up-to-date information. Currently there are thousands of local and personal databases storing key technical and licensing data in a very unsatisfactory manner. An absolutely fundamental need for effective groundwater management and protection is a comprehensive, publicly accessible, groundwater database (GDB). The complete lack of a GDB is seriously constraining the formulation and implementation of effective groundwater management throughout China. The inability to access information, which at times is part of institutional secrecy, encourages inaction or incorrect decisions. GDBs are well established in almost every country where significant groundwater is used. The lack of such a database in China is surprising.*" (MWR, 2001).

Box 1 shows that the levels of uncertainty can remain high even where considerable efforts are made to analyse and interpret raw hydrogeological data.

Drawdown externalities and the sustainability issue

The impacts of overabstraction and water-level declines have been reported widely. It is sufficient to note here that overabstraction can lead to a wide array of social, economic and environmental consequences including:

- critical changes in patterns of groundwater flow to and from adjacent aquifer systems;
- declines in stream base flows, wetlands, etc. with consequent damage to ecosystems and downstream users;
- increased pumping costs and energy usage;
- land subsidence and damage to surface infrastructure;
- reduction in access to water for drinking, irrigation and other uses, particularly for the poor;

Box 1: The significance of groundwater data and its uncertainties

The problems of compiling groundwater data and interpreting abstraction records to establish and model the status of an aquifer system are well illustrated by interim reports on the Northwestern Sahara Aquifer System (SASS/OSS, 2001). This massive system covers Algeria, Tunisia and the Libyan Arab Jamahiriya and broadly comprises two super-imposed sandstones: Continental Terminal and Continental Intercalaire. Up to 8 000 borehole records and associated abstraction data from 1950 to 2000 have been compiled and a regional groundwater model constructed. A preliminary analysis of abstraction data indicates a two to threefold increase in pumped volumes throughout the aquifer system beginning in the late 1970s, peaking in 1990 and thereafter showing stabilization or slight decline. Abstraction from the system as a whole is estimated at 80 m^3/s. In 1950, abstraction was estimated at 13 m^3/s and in 1975 had reached 25 m^3/s. The impact on the overall water balance of the system is being refined through the application of a regional hydrogeological model. However, the distribution of boreholes indicates that the generation of drawdown externalities will be very localized. At control-point observation boreholes, there has been a marked lowering of the piezometric surfaces since 1980, with total drawdowns since 1950 typically of the order of 20-40 m in the exploited aquifer blocks. However, the control on abstraction data is variable (in Tunisia abstraction yearbooks have been published since 1973). In general, the levels of uncertainty associated with the data derived from a variety of data sources from the period of record are manifold. The authors of the reports emphasize the fundamental methodological differences that need to be appreciated when dealing with hydrogeological as opposed to hydrological data; specifically piezometric (water pressure head or level) altitude corrections and methods of analysis and validation in relation to hydrogeological time-series data. For these reasons, the error terms that need to be attached to any hydrogeological observation are significantly higher than those normally associated with surface water data. Thus, even where great effort and thought go into standardizing raw hydrogeological data in preparation for modelling activities, levels of uncertainty will remain high. In this particular case, model results do show a good match with the control observation data and thus provide a broad picture of the aquifer's evolution as development has proceeded. However, the data and model results would probably be too coarse for the establishment of quantitative pumping rights in specific aquifer blocks.

Box 2: The Tripoli Statement

More than 600 participants from more than 20 countries and regional and international organizations and associations attended the International Conference on "Regional Aquifer Systems in Arid Zones – Managing Non-Renewable Resources" in Tripoli, 20-24 November 1999. The Statement of the Conference reads:

"We the Participants of the Conference recognize that:

In most arid countries the scarcity of renewable water supplies implies a serious threat to sustainable coupled and balanced socio-economic growth and environmental protection. This threat is clearly more pronounced in the less wealthy countries. In many arid countries, however, the mining of non-renewable groundwater resources could provide an opportunity and a challenge, and allow water supply sustainability within foreseeable time-frames that can be progressively modified as water related technology advances.

The Conference marks a milestone in the discussion of the emerging concept of planned groundwater mining. We the Participants consider that:

Adoption of this concept at national level could have international repercussions;

A national integrated water policy is essential with, where feasible, priority given to renewable resources, and the use of treated water, including desalinated water.

We recommend that: groundwater mining time-frames should account for both quantity and quality with criteria set for use priorities, and maximum use efficiency, particularly in agriculture; care should be exercised to minimize the detrimental impact to existing communities; consideration should be given to the creation of economical low water consuming activities. We the Participants further consider that in situ development, or development based upon transferred mined groundwater, depend upon many non-hydrogeological factors outside the scope of this Conference. Nevertheless, hydrogeological constraints need to be defined for both planners and the end users. We recommend the participation of the end users in the decision making process and the enhancement of their responsibility through water use education and public awareness. We believe that for efficient water-use, cost recovery could eventually be necessary. In recognition of the fact that: some countries share aquifer systems; international law does not provide comprehensive rules for the management of such systems as yet, and clearly groundwater mining could have implications for shared water bodies; we the Participants draw the attention of Governments and International Organizations to the need for: rules on equitable utilization of shared groundwater resources, prevention of harm to such resources and the environment, exchange of information and data. We also encourage concerned countries to enter into negotiations with a view of reaching agreements on the development, management, and protection of shared groundwater resources."

- increases in the vulnerability of agriculture (and by implication food security) and other uses to climate change or natural climatic fluctuations as the economically accessible buffer stock of groundwater declines.

The term 'overabstraction' should not be confused with the term 'groundwater mining'. The latter term refers only to the depletion of a stock of non-renewable groundwater, leaving the aquifer dewatered indefinitely. The planned mining of an aquifer is a strategic management option if the full physical, social and economic implications are understood and accounted for over time. The bulk of the exploited groundwater in the world's principle stratiform aquifers was emplaced during the last 100 000 years. This applies equally to the coastal aquifers of Europe (Edmunds and Milne, 2001) as it does to the aquifers underlying the arid regions of the world where current recharge is nil or minimal. It is perhaps in the arid zones that most attention has been focused and the 'Tripoli Statement' echoes the concern for aquifers in arid regions (Box 2). A case in point is the current exploitation of the Nubian Sandstone Aquifer System (NSAS), which it is estimated represents 0.01 percent of the estimated total recoverable freshwater volume stored in the NSAS (Box 3).

However, even when recharge is taking place, replenishment by downward percolation of meteoric water shows high interannual variability and is a complex physical process that is difficult to evaluate (Lerner, 1990; Simmers *et al.,* 1992). Therefore, even in actively recharging systems, overabstraction should not be defined in terms of an annual balance of recharge and abstraction. Rather, it needs to be evaluated on an interannual basis since the limit between the non-renewable stock and the stock that is replenished by contemporary recharge from surface percolation is usually unknown. However, what is of real importance to decision-makers and well users is the overall reliability and productivity of a well (in terms of water levels, volumes and water quality) during a given time period. Therefore, if a well taps a particular aquifer, what is its sustainable rate of exploitation given variable periods of recharge and drought? The answer to this question is not trivial, and it requires a certain level of precision in understanding the dynamics of the physical system. Therefore, in practice, the only real management indicator for a community of groundwater users is the maximum admissible drawdown they are prepared to accept.

Yemen presents dramatic evidence of the consequences of overabstraction. According to the recent Water Resource Assessment of Yemen (WRAY-35, 1995): "*...almost all important groundwater systems in Yemen are being overexploited at alarming rates.... Worst-case predictions made in 1985 on possible depletion of the Wajid sandstone aquifer of the Sadha Plain... have unfortunately come true and groundwater levels have declined on average some 40 metres in only nine years.*" High-quality water available in shallow aquifers near Sana'a, Yemen's capital, is expected to be depleted within a few years. This contrasts with rising water levels due to sewage infiltration under the city itself.

The scale and rate of groundwater abstraction are related directly to the massive expansion in pumping capacity that has occurred over the past five decades in many parts of the world. The number of diesel and electrical pumps in India has risen from 87 000 in 1950 to 12.58 million in 1990 (CGWB, 1995) and to an estimated 20 million today.

The impacts of long-term abstraction are readily apparent in regions where spring and seepage zones disappear or where users have to dig or drill deeper to chase a locally falling phreatic or piezometric head. In addition, the aquifer systems themselves are vulnerable to abstraction in many complex and often not immediately apparent ways. As in most discussions concerning groundwater overabstraction, these statistics focus on rates of water-level decline and the degree

Box 3: The Nubian Sandstone Aquifer System

Background

The Nubian Sandstone Aquifer System (NSAS) consists of a number of aquifers laterally and/or vertically interconnected, extending over more than 2 000 000 km² in the east of the Libyan Arab Jamahiriya, Egypt, northeast Chad and north Sudan. The main components of the NSAS include:

- Palaeozoic continental deposits (mainly sandstone);
- Mesozoic continental deposits, pre-Upper Cenomanian (Nubian sandstone sensu stricto);
- Post-Eocene continental deposits (mainly sandstone) in the Libyan Arab Jamahiriya, equivalent to carbonate rocks aquifer in Egypt (this component communicates with the underlying Mesozoic or Palaeozoic aquifers through Mesozoic-Cenozoic low-permeability formations).

The Nubian aquifers including the Palaeozoic and Mesozoic deposits older than the Pre-Upper Cenomanian extend over the whole Nubian Basin, although becoming very saline in the northern part. The Nubian deposits are outcropping or subcropping in all that part of the basin located south of the 26th parallel, in which the aquifer system is under unconfined condition. The unconfined part of the Nubian aquifers includes the most important groundwater potential of the whole basin. The extension of the cones of depression resulting from the water abstraction in existing and planned well fields in that part of the Nubian domain is always limited and makes it possible to multiply the centres of extraction.

The Post-Nubian aquifer, corresponding to the Post-Eocene deposits, occurs only in Egypt and the Libyan Arab Jamahiriya. It is more important in the Libyan Arab Jamahiriya in term of development potential.

The aquifer systems are not in equilibrium. The observed groundwater flow from south to north and the present natural outflow in the large and deep evaporative areas between Ajdabyia and Cairo are not due to present recharge but to paleo-recharge gradients established during pluvial periods of the late Quaternary.

Water resources and beneficial uses thereof

Data collected in the framework of the programme funded by the International Fund for Agricultural Development (IFAD) on the NSAS made it possible to estimate the amount of freshwater stored in the two aquifer systems. Following the results of this study, it is possible to envisage diverse scenarios considering different options for the development of water resources. The following table presents the main results of this freshwater resources assessment:

Country	Nubian system (Palaeozoic and Mesozoic sandstone aquifers)		Post-Nubian system (Miocene aquifers)		Total freshwater in storage[1]	Total recoverable groundwater[2]	Present extraction from systems		
	Area	Freshwater storage	Area	Freshwater storage			Post-Nubian	Nubian	Total
	(km²)	(km³)	(km²)	(km³)	(km³)	(km³)	(km³)	(km³)	(km³)
Egypt	815 670	154 720	426 480	97 490	252 210	5 180	0.306	0.200	0.506
Libya	754 088	136 550	494 040	71 730	208 280	5 920	0.264	0.567	0.831
Chad	232 980	47 810	n.a.	n.a.	47 810	1 630	n.a.	0.000	0.000
Sudan	373 100	33 880	n.a.	n.a.	33 880	2 610	n.a.	0.840[(3)]	0.833
Total	2 175 838	372 960	920 520	169 220	542 180	15 340	0.570	1.607	2.170

[1] Assuming a storativity of 10^{-4} for the confined part of the aquifers and 7% effective porosity for the unconfined part.
[2] Assuming a maximum allowed water level decline of 100 m in the unconfined aquifer areas and 200 m in the confined aquifer areas.
[3] Most extracted in the Nile Nubian Basin (833 Mm³/year) which is not considered to be part of the Nubian Basin.

Source: CEDARE/IFAD Programme for the development of a Regional Strategy for the Utilisation of the Nubian Sandstone Aquifer System.

Most of the water extracted from the NSAS is used for agriculture, either for large development projects in the Libyan Arab Jamahiriya or for private farms located in traditional oases in Egypt (New Valley). However, an important project designed for transporting water to the coast from the NSAS is under development in the Libyan Arab Jamahiriya. The project is already supplying some 70 Mm³/year of water to Benghazi and to the major coastal cities west of Ajdabyia. From the above figures, it appears that the present extraction represents only 0.01 percent of the estimated total recoverable freshwater volume stored in the NSAS.

Significant issues concerning the NSAS

The large groundwater development projects planned in southern Egypt and the Libyan Arab Jamahiriya within the Nubian Basin are not expected to induce any significant effect beyond the common border between the two countries. The different options of water resources development can influence the amplitude of the cone of depression, such as the one that may be extended beyond the Egyptian-Sudanese border over some 50-70 km, which could be expected to be generated if particularly intensive water extraction were realized in southwest Egypt. In the north, the groundwater development at Siwa oasis from the deep aquifer (Nubian) is close to the freshwater-saltwater interface. Increasing the present abstraction may draw saline water into the freshwater aquifer. The development of a well field in the Jaghbub area, located in the Libyan Arab Jamahiriya in a symmetric position to Siwa with respect to the border line, would probably augment the risk of deterioration of the water quality in the Nubian aquifer.

An IFAD-funded project (for the development of a regional strategy for the utilization of the NSAS, operated by CEDARE) is assessing the regional implications of various groundwater development scenarios based on the NSAS. It will propose consultation mechanisms for the joint management of the water resources including systematic data exchange on groundwater extraction and on water-level and water-quality fluctuations.

to which extraction estimates exceed replenishment estimates. The provenance of the replenishment, whether recharge from the surface or leakage from adjacent aquifers is rarely known with any precision. However, sustainability is defined implicitly as a level at which draft and recharge are balanced (the 'sustainable yield' of an aquifer). This assumes that a steady state can be achieved in which water levels are stabilized. This narrow focus is often misleading. Pumping will induce water-level declines regardless of whether or not the 'sustainable yield' of an aquifer has been exceeded. These initial water-level declines can have major social, economic and environmental impacts long before sustainability of the groundwater resource base is threatened in any quantitative sense.

Discussions of groundwater sustainability need to focus on the ability of the resource to produce key services (including environmental services) and on the economic costs and impacts on equitable access that the loss of such services would entail. For example, declining water levels generally have large equity impacts particularly in the developing world. Wells established for drinking supply often go dry, forcing women and children to walk long distances or wait in line to obtain water to meet domestic needs. Less wealthy farmers are often only able to afford shallow, low-capacity wells. As water levels drop, they can be excluded progressively from access to groundwater by the costs of well deepening and new equipment. This effect undermines the food security and economic development benefits generated by access to groundwater. Water-level declines increase considerably the probability of environmental impacts on streams, wetlands and the occurrence of subsidence. They also increase the probability that low-quality water and pollutants will migrate into key freshwater aquifers. Finally, water-level declines can lead to economic exhaustion of the replenishable groundwater resources. As levels decline, drilling and pumping costs increase. Water may still be physically available, but the cost of extraction can be sufficiently high to exclude all but the highest-return applications. The case of the Gangetic Basin in India and Bangladesh illustrates many of these issues (Box 4).

As the example on the Gangetic Basin (Box 4) illustrates, overdraft and water-level declines typically affect the sustainability of uses that are dependent on groundwater long before the replenishable resource base is threatened with physical exhaustion. Therefore, the sustainability of socio-economic activities in relation to falling groundwater levels is complex. Any analysis of this particular issue needs to examine carefully a range of consequences, including:

- Protection of drinking-water supply sources (both access and quality).
- Equity in access and allocation and poverty alleviation.
- Maintenance of environmental values dependent on groundwater levels or groundwater discharge to watercourses.
- Food security and agricultural production.
- Economic development.

Box 4: Groundwater development in the Gangetic Basin

The Gangetic Basin is filled with unconsolidated alluvial materials to a depth of about 6 000 m and receives an average of 1 500 mm of rainfall each year (Rogers, Lydon *et al.*, 1989). The total amount of groundwater in storage would be sufficient to meet all needs for centuries with little danger of physical depletion.

Despite the large stock of water in storage, groundwater development is having major impacts. In Bangladesh, water-level fluctuations are causing shallow wells to go dry, particularly during summer. This creates major difficulties for villagers in obtaining drinking-water supplies (Sadeque, 1996). It also has major equity effects. Wealthy farmers can afford to deepen existing wells or install new ones; those who are less wealthy often cannot afford the cost of chasing the water table. Environmental impacts could also be major. Kahnert and Levine (1989) commented in their summary of a symposium on groundwater irrigation and the rural poor that: "the data show significant seasonal variations in both the water table and the flow of the Ganges in its lower reaches." Participants in the symposium also expressed concern about the potential impact of increased groundwater extractions on the base flow into the Ganges River at low flow periods. Modelling activities currently underway appear to substantiate these concerns. Results suggest that dry season flows at Farakha Barrage near the Bangladesh border could decline by about 75 percent if historical groundwater development patterns continue (Ilich, 1996).

Declines in dry season flow are a point of contention between India and Bangladesh. For Bangladesh, these flows are critical for irrigation, drinking-water supply and for sustaining mangrove areas along the coast. Furthermore, in Bangladesh, 70-80 percent of total animal protein consumption is dependent on fish. Activities affecting floods and drainage could interrupt approximately 60 percent of the nation's fish production (Rogers, Lydon *et al.*, 1989).

Finally, declining water tables have major implications for energy consumption. India's electricity deficit now runs at 19 percent for base load and at more than 30 percent for peak load. Much of the problem relates to agricultural use, primarily pumping for irrigation. In many states, official figures published by the state electricity boards indicate that agricultural demand exceeds 40 percent of consumption. In some, such as Haryana, it exceeds 50 percent. While these figures may be misleading (they include massive 'non-technical' losses), the rate of growth in agricultural electricity consumption has been dramatic. Power for groundwater pumping is highly subsidized. Farmers usually pay a flat rate based on pump horsepower. As a result, when water tables fall, farmers have little incentive to reduce extraction. This exacerbates both energy and overdraft-related problems.

Groundwater extraction and migration of low-quality water

Vulnerability to declines in groundwater quality as a result of increased groundwater extraction is particularly high in certain contexts. These include:

- Coastal zones: Intrusion of saline ocean water is a common result of pumping, particularly in locations where sediments are highly permeable and in small islands and atolls.
- Interbedded high- and low-quality aquifers: In many locations, aquifers containing high- and low-quality water are interlayered.
- Locations where low-quality water is present on the surface or in adjacent rock formations. Pumping often causes lateral migration of low-quality water from adjacent aquifers.
- Locations where rock formations encourage rapid flow. Water flows much more rapidly through karstic limestone or other rock formations where large interconnected fractures or cavities are present. These locations tend to be much more vulnerable to rapid contamination from chemical and bacterial sources.
- Locations where the geochemistry of adjacent waters and/or the geological formations is incompatible. Groundwater geochemistry often differs. This can result in a wide variety of chemical reactions when water containing different levels of key constituents or having differing pH or redox potentials is drawn into and mixes with water in pumped aquifers.

Several of the above characteristics are often present at a single location. However, beyond the vulnerability of different regions, it is important to recognize that quality deterioration is an unavoidable result of use.

The impact of quality declines on the sustainability of uses that are dependent on groundwater can equal or exceed the direct impact of groundwater overdraft. Quality declines can reduce and destroy the value of entire aquifers as a source of water. To this extent, quality declines associated with increased groundwater extraction can have as major an impact on the sustainability of key uses as more publicized problems such as groundwater overdraft.

Rising water levels and waterlogging

Waterlogging induced by irrigation

Pakistan and India contain some of the most extensively documented cases of irrigation-induced waterlogging and salinization. However, even here, it is difficult to evaluate the extent of problems based on available figures. In India, the Ministry of Agriculture estimated that the total area affected by waterlogging as a result of both groundwater rises and poorly controlled irrigation was 8.5 million ha in 1990 (Vaidyanathan, 1994). In contrast, estimates made by the Central Water Commission for 1990, which considered only areas affected by groundwater rises, totalled 1.6 million ha (Vaidyanathan, 1994). Regardless of the actual extent, waterlogging problems represent a major surface water and groundwater management challenge. This challenge cannot be addressed in the absence of an integrated approach that incorporates surface-water imports and use as well as groundwater. Large areas in Pakistan face similar problems. Rising water levels in the command of surface irrigation systems have fundamental implications for the sustainability of social objectives that are groundwater dependent. In the case of food security, estimates indicate that irrigation-induced salinity and waterlogging reduce crop yields in Pakistan and Egypt by 30 percent (FAO, 1997). In India, the problem is serious enough to threaten the growth of the agricultural economy (Joshi *et al.*, 1995). The impact of waterlogging and salinization on farmers and regional economies can be insidious. In the initial years, the introduction of irrigation often causes a dynamic transformation of regional and household economies. Farmers introduce high-yielding varieties of grain and are able to grow valuable market crops. Wealth is created. However, as the water table rises, the 'bubble economy' based on unsustainable water management practices deflates. Once salinized, land and the unsaturated zone of the soil are difficult and expensive to reclaim. Ultimately, many farm families (and regional economies) may be worse off than before the introduction of irrigation unless sustainable and affordable methods of remediation are found.

Water-level rises under urban areas

Water-level rises are a major feature in many urban areas, particularly once cities begin to rely on imported supplies. Although urbanization may reduce direct infiltration of rainfall because of the large impermeable area created, recharge below cities is often far higher than pre-urban levels (Morris *et al.*, 1994; BGS, 1995). In a recent study, the increases in recharge under Merida, Hat Yai and Santa Cruz (cities in Mexico, Thailand and Bolivia, respectively) ranged from 130 to 600 percent. In Lima, Peru, recharge has increased from essentially zero to 700 mm/year (Morris *et al.*, 1994). As most of this recharge comes from leaking sewers and water mains, the potential for pollution is high. Where water imports induce rising water levels in unconfined aquifers, the effect enables shallow wells to serve as a major source of water supply for the poor. However, as pollution levels are generally much higher in shallow urban aquifers, particularly in areas not served by sewer systems, those dependent on shallow wells face major health risks. This is well illustrated by the case of Sana'a, Yemen, where water levels under the city are rising despite general conditions of overdraft in aquifers supplying the city. Furthermore,

Box 5: Groundwater management in the Indus Basin

Recent estimates are that irrigated land furnishes 90 percent by value of Pakistan's agricultural production, accounting for 26 percent of its gross domestic product (GDP) and employing 54 percent of the labour force. On the evidence of current investments, priorities in irrigated agriculture outweigh other interventions. Maintaining a bank of soil resources and flow of water resources to support food production to a population growing at 3.0 percent/year has become an imperative for Pakistan. The bulk of this productivity is associated with the Indus Basin.

The Indus Basin is filled with thick alluvial sediments deposited by the Indus River and its five main tributaries (Jhelum, Chenab, Ravi, Sutlej and Beas) forming a thick set (300-500 m) of unconfined and leaky aquifers. Before the introduction of a weir-controlled canal irrigation system, the groundwater table was relatively deep under most of the plain. As a result of the additional recharge introduced by irrigation, the water table started rising at a rate of 15-75 cm/year. The position of the water table before and after the introduction of the large canal networks in the upper part of the basin rose 20-30 m in 80-100 years. The quality of groundwater varies in vertical and horizontal directions and is related to recharge of the aquifer. In general, water from shallow wells located near sources of recharge is of good quality. Along the rivers and in the upper reaches of the doabs, where precipitation is a major source of recharge and maximum canal supply is available, groundwater usually contains less than 1 000 ppm of dissolved solids (1.56 dS/n).

The Indus Basin was developed through surface irrigation in the late nineteenth century but the threat to the system of saline accumulation in irrigated soils was appreciated by the original design engineers. The results have been:

- Public tubewell development started in the 1960s through Salinity Control and Reclamation Projects (SCARPs). Because drainage projects alone generally have a low economic rate of return, priority has been given to locating SCARPs in areas of usable-quality groundwater. As a result, 90 percent of the SCARP tubewells and 95 percent of the pumped groundwater is from freshwater groundwater zones. SCARPs have evolved into groundwater supply projects in which drainage is a by-product.
- The SCARP development has triggered the capacity of the private sector to develop good-quality groundwater (something not appreciated in the early planning stages).
- The salt balances of the Indus Basin and its associated sub-basins have been disrupted as the hydrochemical systems have become progressively closed and the supplemental generation of salt through waterlogging has further exacerbated the positive salt balance.
- The Indus Basin is effectively a saline sink with minimal flushing and outflow. This applies to the Indus Plain as much as to the North West Frontier Province (NWFP) and Baluchistan sub-basins, which are also in danger of becoming closed subsystems.
- The recently launched National Drainage Programme has dropped subsidies from public tubewells in fresh groundwater areas.

The physical and chemical environment in which groundwater is found and is evolving is complex, particularly in the shallow horizons that have experienced recent groundwater recovery and quality changes. Relatively fresh groundwater occurs side by side with saline groundwater or under or overlain with saline groundwater. This requires a high degree of operational knowledge in the management of groundwater in order to ensure its sustainability in terms of quantity and quality. Therefore, the identification of hydrogeological processes and the establishment of a physiographic framework are imperative in order to both explain and quantify the groundwater occurrences and the rate of aquifer replenishment and depletion.

high water levels under urban areas cause drainage problems, leading to the creation of stagnant and highly polluted surface water bodies.

Water-level changes in response to vegetation cover

Land-use changes can have a significant impact on groundwater levels. Forest and vegetation cover have long been recognized as major factors influencing runoff, infiltration and evapotranspiration from shallow water tables. Watershed treatment involving the establishment of tree, bush and other plant cover is widely used as a way of reducing runoff and increasing infiltration. This is frequently assumed to increase recharge and is advocated as a core part of packages to address groundwater overdraft. However, the effect of surface vegetation on groundwater levels is not automatic. It depends on the balance between improvements in infiltration caused by increased vegetation and relative changes induced in evapotranspiration. In some cases, removal of forest cover has caused water levels to rise significantly with major environmental consequences, e.g. in much of New South Wales, Australia.

Pollution externalities

Pollution is widely recognized as one of the most serious challenge to the sustainable management of groundwater resources. The significance of pollution for groundwater resources is increased by the long time scale at which processes affecting groundwater function. As Morris *et al.* (1994) comment: "It is important to appreciate the differences between surface water and groundwater systems. In the former, the water is typically being replenished, at least in the case of rivers, within time-scales of weeks or at most months. Replenishment times for groundwater systems are very much longer. This is because water usually takes many years to move through the soil and unsaturated zone of the aquifer. Once there, it can take a further period of many tens or hundreds of years to flow into a supply borehole." In some of the deeper aquifers, groundwater is likely to be thousands of years old (Edmunds and Wright, 1979; Edmunds *et al.*, 1987). In addition to the relatively slow movement of water in many aquifers, rocks and soil absorb and otherwise attenuate the presence of pollutants. Not all aquifers are equally vulnerable to pollution. Those where fractures or cavities permit rapid flow tend to be more vulnerable than those where water flows slowly through porous media and more opportunities exist for attenuation of pollutants. However, vulnerability to pollution has an inverse relationship to the difficulty of remediation. Once polluted, slow movement of groundwater through a porous aquifer generally makes cleanup difficult, expensive, and in some cases impossible.

Beyond the inherent vulnerability of aquifers to contamination, much depends on the nature of pollutant sources. Contaminant behaviour varies greatly with respect to the specific transport properties in each aquifer system. In addition, the range of contaminant types is increasing as new products appear in effluent disposal and land application. Three main sources of groundwater pollution are: agricultural, urban and industrial.

Agricultural pollution

In many developing countries, agricultural chemical use has been low in comparison to levels in industrialized countries. This may no longer be the case, particularly in countries such as India and China where irrigation is extensive. Concerns over groundwater pollution from agricultural chemicals were raised as a major issue in India more than two decades ago (Chaturvedi, 1976) but few data were available. At that time, the level of agricultural chemical use was very low. However, by 1991, fertilizer use per hectare of agricultural land was 60 percent higher than in the United States of America (Repetto, 1994). At present, no agency in India has a systematic programme for monitoring potential non-point sources of pollution. However, fragmentary data indicating the potential extent of agricultural pollution problems are available. For example, maps prepared by the Central Ground Water Board (CGWB) show nitrate concentrations in Gujarat exceeding 45 mg/litre (the WHO's recommended maximum for drinking-water) in more than 370 sample sites scattered across the state (Phadtare, 1998). How much of this pollution is related to agricultural pollution and how much to domestic or other sources is unknown.

Aside from non-point-source considerations, it is important to recognize that nitrate and other nutrient pollution in groundwater is often related to agricultural practices other than the use of chemical fertilizers. Any location where animal wastes are concentrated, such as feed lots or poultry farms, can release high levels of nutrients into groundwater. In addition to nutrients, pesticides and herbicides are other major sources of groundwater pollution related to agriculture. In some circumstances, soils can absorb or immobilize a large fraction of such agricultural chemicals. However, many pesticides and herbicides break down slowly under aquifer conditions or can transform into more toxic compounds. As a result, they can persist over long time periods. In any case, groundwater pollution data are generally scarce and chemical analysis of water samples needs to be specific to detect their presence.

The dispersed nature of sources of pollutants is a core challenge facing both monitoring and control of groundwater pollution related to agriculture. Unlike industry or municipal sewage systems, agricultural pollutants are dispersed over large land areas. While return flows in drainage canals can be monitored, it is difficult to determine the extent of direct seepage of pollutants through soils and into the groundwater until contaminant concentrations in groundwater become significant.

Urban groundwater pollution

The additional recharge in urban areas is derived principally from leaking sewers and other wastewater sources. Broken sewers in the United States of America are estimated to lose 950 Mm^3 of wastewater each year (Pedley and Howard, 1997). Much of this represents polluted recharge to groundwater. Direct leakage of wastewater to groundwater in developing countries is probably much higher. In many cities, a large portion of the wastewater generated is discharged directly into unlined canals. Where sewer systems exist, leakage levels are almost certainly much higher than in the United States of America because of lack of resources for maintenance, variability in construction materials and absence of adequate treatment facilities. Furthermore, in many urban and peri-urban areas, pit latrines and soak pits are used to dispose of domestic wastewater. These are often relatively deep (more than 3 m) and discharge wastes below the soil and weathered zone layers that have the greatest capacity to filter, absorb and otherwise attenuate pollutant concentrations (Pedley and Howard, 1997).

The impact of urban wastewater discharges on groundwater is well illustrated by the cases of Santa Cruz, Bolivia, and Hat Yai, Thailand. In both these cities, direct discharge of untreated wastewater has led to substantial increases in pollutants (NO_3^-, NH_4^+, Cl^-, faecal coliforms, and dissolved organic carbon) in the shallow aquifers. The quality of deeper groundwater is still good but pollution fronts are moving downward in response to extraction from deeper levels for drinking-water supply and other uses (Morris *et al.*, 1994). This situation is typical of many cities, particularly in rapidly urbanizing sections of the developing world.

Water supply officials tend to recognize the potential impact of waste discharge on chemical contamination of groundwater by nitrates and other compounds. However, it is often assumed that the filtering action of aquifers and relatively long residence times underground are sufficient to remove pathogens except where open or poorly sealed wells are contaminated directly by surface water inflows. This perception is inaccurate. According to Pedley and Howard (1997): "Bacteria can survive up to 50 days or more in subsurface environments and viruses for far longer."

Overall, the pollution of shallow aquifers under cities represents a major threat to the sustainability of drinking-water supplies in many urban areas throughout the world (BGS, 1995; World Bank, 1998). This threat is particularly high where regional hydrogeological conditions permit rapid flow of contaminated water into aquifers and the wells tapping them. For example, aquifers in karstic carbonate rocks or fracture zones are far more susceptible to contamination than aquifers where groundwater flows through porous media such as soil or sandstone. The threat is also particularly high where large portions of the urban population both dispose of untreated wastes directly through soakaways and latrine pits and also depend on shallow wells for drinking-water supply.

Industrial pollutants

Public attention with regard to groundwater pollution often focuses on 'hot spots' where industrial activities have polluted large areas. Sites of this type often receive national attention. Jetpur, a textile town in Gujarat, India, where more than 1 200 small industrial units drain effluents containing cadmium, zinc, mercury, chromium and other pollutants into small rivers and thence into groundwater, is a prime example (Moench and Matzger, 1994). Governmental monitoring and cleanup activities also tend to focus on high-profile sites. The 'superfund' sites in the United States of America and the activities of the state and central pollution control boards in India are typical of many governmental initiatives, particularly during early phases, when the significance of groundwater pollution is only beginning to be recognized. In India, the Central Pollution Control Board has a programme to monitor groundwater quality in 22 critically polluted sites (Moench, 1996). However, there is no baseline monitoring of potential industrial pollutants except within these hot spots.

The hot-spot focus of public attention and many government initiatives tends to downplay the importance of dispersed sources of industrial pollutants such as trace metals and organic solvents. Because of their low solubility, many such pollutants have extremely long residence times in aquifers. Because they do not dissolve rapidly, they can remain indefinitely as a concentrated source of pollution within an aquifer. In some cases, gradual volatilization of organic solvents in aquifers can become an air-quality hazard. Dispersed sources of industrial pollutants are much harder to identify, monitor and control than the effluent from specific factories or industrial areas. As such, these dispersed sources may well represent a greater threat to groundwater resources than concentrated industrial effluent flows.

Data on groundwater pollution in developing countries are generally unavailable. This is particularly the case for pollution related to dispersed sources such as mining activities, underground storage tanks and direct discharge of effluent to water bodies and watercourses. However, with increases in transportation needs and industrial activities, the number of sites where pollution is occurring is increasing rapidly.

Implications of groundwater pollution

The full impacts of groundwater pollution on health, agriculture and the environment have not been assessed comprehensively. In the case of health impacts, Pedley and Howard (1997) observe that: "*The contribution made by contaminated groundwater to the global incidence of waterborne disease cannot be assessed easily; for many countries the incidence of waterborne disease is not known accurately and the data for groundwater usage are not available. Where public health statistics are available, the data are insufficient to determine the source of the water involved in the transmission of the disease.*" However, in comparison with other topics such as the environment, collection of public health data is widespread and relatively well established. The difficulty of assessing the impact of groundwater changes on health (where at least some data are available) gives an indication of the magnitude of the challenge in assessing impacts on other values.

Lack of information on the health and other costs associated with groundwater pollution and quality declines may lead to questions regarding the importance of these problems. While the dangers of pathogenic organisms are recognized, officials in developing-country situations often emphasize informally the lack of evidence that diseases, such as methoglobanemia, or responses to toxic substances are occurring in any but the most polluted areas. Based on this perception, they often advocate relaxation of standards. Part of this response may be due to the widespread incidence of many other health and disease problems, making diagnosis difficult. Part may also be because of priorities. Pollution control and aquifer remediation are expensive. In the United States of America, substantial debate has emerged over the cost of cleanup in relation to the value of groundwater resources (National Research Council, 1997). In developing countries, demands on limited financial resources are often more intensive. As a result, more questions arise regarding large investments in pollution control or aquifer remediation that have few immediately observable returns.

The above observations do not imply that the human and environmental burdens associated with groundwater pollution are minor. Diseases related to water pollution are a major concern in many parts of the world. In 1994, cholera caused more than 10 000 deaths; in recent years, 25 000 deaths have been caused by typhoid, 110 000 by amoebae; and diarrhoeal diseases have claimed the lives of 3.2 million children under 5 years old (Pedley and Howard, 1997). Comprehensive data on deaths and disease caused by absorption of trace metals and other pollutants are not available. However, overall, days lost to disease and the continuing burden of sickness on society far exceed the actual number of deaths. Although the amount of death and disease that can be attributed to groundwater pollution per se as opposed to surface water pollution is unknown, it adds a continuing burden to the health of large populations, particularly in developing countries. In a similar but mostly undocumented manner, groundwater pollution affects a wide range of other key environmental and social values.

Soil and groundwater contamination from industrial and population expansion is of widespread concern. The prevalence of contaminants at hazardous waste sites is well documented. If they are not removed or sequestered, they can contaminate millions of litres of groundwater over time scales of decades and centuries. The remediation of polluted groundwater is driven by the need to reduce risks by achieving regulatory compliance, or in reducing liabilities, at the lowest cost. The high costs and ineffectiveness of groundwater extraction methods in removing unwanted chemical constituents completely from aquifer systems has motivated the development, testing, and application of *in situ* treatment methods.

Chapter 3
Matching patterns and intensity of use and the resource base

The resource base

A prime focus of UNESCO's initiatives under the groundwater component of the IHP has been on hydrogeological information, research and analysis and their role in policy, public dialogue and integrated water resource management (including risk analysis and transboundary aspects). In particular, the collaboration with the International Association of Hydrogeologists (IAH) has yielded authoritative work in the areas of:

- mapping and modelling of large-scale aquifer systems;
- the nature and variability and recharge styles;
- the scale and intensity of groundwater degradation;
- techniques for enhancing recharge.

The only systematic regional surveys of groundwater occurrences were compiled in the 1980s by the then UNDTCD (now merged into the UNDESA) under the Water Series publications. This comprehensive work has not been updated, but at the time it represented the only systematic portrait of groundwater occurrences on a country-by-country basis with an indication (where possible) of trends in groundwater resource use.

The degree to which the broad array of information on specific aquifers, systems and scientific themes related to groundwater can be combined to present a coherent 'state of the art' on the resource base is questionable. The proliferation of relevant material at country level would overwhelm a global exercise and the benefits would be negligible in comparison with the urgent local problems on which groundwater management has to focus.

Patterns and intensity of groundwater use

The specific role of groundwater in socio-economic development has been widely recognized for the past 50 years (United Nations, 1960 and 2000). The specific advantage of groundwater is its ubiquity, and this ubiquity presents particular problems in its management. The spatial and temporal distribution of groundwater occurrences has presented certain opportunities for the development of rural communities, urban centres and irrigated agriculture. It is the patterns of groundwater use that serve as a starting point for considering management options. For example, the patterns and management of groundwater and aquifer use in urban areas are distinguishable from those in rural areas. There are two distinct styles of use that exhibit and require distinct management solutions for each setting. In many arid and semi-arid urban areas, local aquifers are often the water resource of last resort and also the ultimate pollution sink. In these areas, the range of services provided by underlying aquifer systems is usually more complex than that demanded by adjacent rural dwellers. The systems of rights in use differ markedly. Rural users generally abstract groundwater themselves through wells that they own and control. Urban users are one step removed with extraction being the responsibility of organizations such as municipal supply utilities. This difference has legal implications. Rural users anticipate access in the form of direct abstraction from local aquifers (irrespective of their legal or customary status). It is important that this access be protected in the form of a property right or a declared right in use. On the other hand, most urban dwellers and businesses anticipate municipal services derived from groundwater resources without any sense of real engagement with the resource (or right in use).

The rural pattern is characterized by a mix of dispersed potable, stock-watering and irrigation applications. The irrigation can often be locally intensive and crowd out other uses. Drawdown externalities are evident in many systems as shallow wells dry up and pressures are lost. The urban pattern differs markedly as the aquifer systems are expected to provide reliable sources of potable water from dedicated well fields (and supplementary private sources where reticulated systems are unreliable) and also to absorb effluent. Groundwater rise in superficial aquifers underlying urban areas poses geotechnical and public health problems

Irrigated agriculture and municipalities are often perceived as the predominant users of groundwater; urban centres looking for low to moderate volumes of high-quality water, and irrigated agriculture looking for high volumes of raw water, being indifferent about quality unless the groundwater is saline.

In practice, urban centres are struggling to maintain the quality of local groundwater sources and urban development is expanding on recharge areas. Irrigated agriculture is using large volumes of high-quality groundwater in inefficient field distribution and drainage systems with attendant problems of salinity and waterlogging. The advent of mechanized boreholes has encouraged rapid depletion of deeper, non-renewable sources. While this pattern of groundwater use is unsustainable in terms of renewable water resources, planned depletion and the assumption of substitutionality (Schiffler, 1998) are seen as valid development paths.

What is much more difficult to resolve is the sustainable development of the apparently low-intensity use of shallow groundwater in rural communities. In fact, this use is highly intensive in terms of the available resource base. In many semi-arid zones of the world, the locally exploitable groundwater resource is often depleted or drained by the end of the dry season. In addition, domestic use is not limited to potable water and personal hygiene use, but also to stock watering and small-scale irrigation. In aggregate, these uses are significant and key to the welfare and development of the bulk of the populations in developing countries.

There is now increasing pressure for the specific role of groundwater in environmental systems to be realized and accounted for. To this extent, environmental systems could be counted as a 'user'. The problem lies in finding a way to allocate between uses and the need to value the set of environmental benefits and compare them with extractive uses. This is not a trivial issue, and a range of standard cost-benefit analysis and multicriteria analyses have been deployed to try and reconcile these competing uses.

Trends in groundwater research and development

The bulk of hydrogeological and groundwater development/protection research remains largely technocratic. The volume of social research related to groundwater use is small in comparison with the volume of research on groundwater flow and remediation. This trend is changing to the extent that the 2002 IAH Congress focused specifically on the human uses of groundwater (the 2001 Congress dealt with new approaches to characterizing groundwater flow). Box 6 shows how use issues are also being recognized increasingly within UNESCO's IHP VI.

Box 6: Implementation plan of the groundwater component of UNESCO's IHP VI

1. Establishment of guidelines for transboundary aquifer resources management.
2. Improvement of the management of shared river basins through understanding groundwater interactions.
3. Establishment of an international groundwater resources assessment centre.
4. Effects of global changes on groundwater recharge, especially in arid and semi-arid regions in relation to water resources management.
5. Methodologies for risk assessment of wastewater reuse on groundwater quality.
6. Development of methodology for studying responses of aquifers to extreme hydrological events.
7. Study of the dynamics of groundwater flow and chemistry in closed basins including long-term effects, especially in arid zones.
8. Guidelines for delineation of protection zones around public groundwater supplies and management policy.
9. Development of groundwater policy and management for wetlands protection and biodiversity conservation.
10. Evaluation of the impact of land-based sources of pollution on coastal zone resources.
11. Methodology for enhancing communication between water specialists, decision-makers and communities to strengthen public participation in groundwater protection.

In addition, two cross-cutting programme components FRIEND and HELP interact through their operational concept with IHP VI and will support any joint UN-agency programmes. Other specific initiatives on groundwater quality protection and remediation will require the application of innovative and especially cost-effective methods to clean up pollution in soils and aquifers including:

- *in situ* bioremediation approaches;
- reactive barriers;
- source zone treatment;
- flow and transport modelling;
- site characterization.

These activities respond to declared needs from the specialist groups to examine institutional and policy issues in tandem with the technical matters.

However, there appears to be a reluctance to view groundwater resource management from the perspective of the de facto regulator, i.e. the individual user with a mechanized pump. This suggests a need to approach groundwater management as a socio-economic issue together with a sound technical perspective. The institutions 'responsible' for managing and regulating groundwater resources need to focus on social mobilization as a priority. This does not imply that the role of technical understanding of aquifer systems is reduced. Rather, it strengthens the case for enhanced understanding and communication of these systems within broader political and socio-economic frameworks. However, the fact remains that aquifer systems are difficult (and expensive) to characterize, monitor and regulate.

The work by Shah (1993) is one of the few attempts to analyse the political and socio-economic context of groundwater. However, the distributed nature of the resource necessitates managing a large number of diffuse, low-intensity investment decisions. If these investments in groundwater are to be realized, then it is contingent upon the investors to examine the sustainability of the resource and the regulation of its abstraction by the multiplicity of users.

Therefore, project formulators need to be aware of the potential problems associated with groundwater development. In many cases, these have less to do with volumes of abstracted water than with the changes in groundwater levels that the abstraction produces. Thus, resource assessments need to focus on establishing the local hydrodynamic limits of abstraction and waste disposal. Equally important is the understanding of the potential for managing or influencing the new patterns of use, patterns that are often highly dispersed and individualized (Burke and Moench, 2000).

Chapter 4
The socialization of groundwater issues

Introduction

The promotion or permission of groundwater development (by design or default) has been successful. The uptake of groundwater use in preference to surface water supplies of raw water is apparent in irrigated agriculture (where the reliability and flexibility confer enormous advantage) and also in municipal supplies where quality is more consistent. However, the aggregate impact of millions of individual pumping decisions, while highly conditioned by the hydrogeological status of the pumped aquifers, is evident in falling groundwater tables and declining water quality. Traditional water management approaches may not have been able to 'socialize' the responses required in order to reduce drawdown and pollution externalities across communities of groundwater users.

Impacts and responses

The impact of emerging groundwater problems can be intensely local or regional in effect. However, the types of management responses to address the particular circumstances may not have been adjusted to these patterns of use and impact. In particular:

- Large-scale, publicly funded tubewell development has tended to be supply driven.
- Legal and regulatory provisions at national level cannot be policed adequately.
- Forcing the relaxation of overexploited systems may be impossible even where the economic rationale is compelling.
- Regulation of the drilling industry may provide information but may also inhibit development in some circumstances.
- The enhancement of indirect recharge may work for shallow groundwater circulation, but recovery of deeper systems requires sophisticated injection and alternative sources of high-quality water.

Progressive attempts to encourage the formation of groundwater user associations are in their infancy in developing countries. The results of irrigation management transfer and water user associations are now being documented, but specific groundwater examples have yet to emerge in the literature. Large-scale initiatives such as those developed for the Ogallala aquifer in the High Plains in the United States of America may have achieved some impact in attenuating the rate of decline of regional water tables. However, they have come too late to keep the agricultural systems productive for more than two or three generations (White and Kromm, 1995).

Social perspective on technical information

The perception and understanding of hydrogeological processes among groundwater users appear to vary considerably (White and Kromm, 1995; Shah, 1993,). Equally, the levels of information required in order to prompt adequate management responses will vary with: the nature of the aquifer (deep conservative or shallow 'flashy' systems); the level of technology applied (handpumps or high-capacity shaft-driven pumps); and the nature of the media (handdrawn maps or Web-accessed data). In many cases, a simple message about overabstraction or pollution may be sufficient to prompt a collective response by the community of groundwater users, provided it is the right information. In the case of the Ogallala, the initial public perception about recharge of the aquifer was erroneous. Only when it was too late did a more realistic perception plant itself in the minds of the users (White and Kromm, 1995). Local knowledge about groundwater-level fluctuations in shallow systems is often sophisticated and involves customary regulation in times of dry-season, water-table recession. For example, such behaviour occurs with the linear alluvial aquifer in Yemen and the more expansive basalt aquifers in Eritrea. Under these circumstances, although the social perspective may be realistic, it will not necessarily inhibit overabstraction. With shallow circulation systems, pumping local aquifers intensively before dry-season recession sets in may be a sustainable strategy in terms of local food security. However, this paradox begs several questions concerning:

- the level of precision required to understand and communicate hydrogeological processes;
- the degree to which patterns and intensity of resource use can be assessed;
- the degree to which risk can be assessed, given the levels of uncertainty and limitations of available data;
- whether current data lend insights into key questions of access and system dynamics.

It is not possible to answer all these questions. Much of the hydrogeological data that is currently collected has little real meaning in relation to the core issues facing users. The level data do not give much indication as to whether users have access to a reliable source of water supply. In addition, they are unlikely to provide much insight into the flow dynamics necessary to really 'manage' an aquifer.

Coping strategies

The need to find practical management practices would suggest that it is important to examine: (i) people's coping (adaptive) strategies, i.e. what populations do when faced with groundwater scarcity problems (water harvesting, alternative livelihoods and demographic shifts); and (ii) the policy implications (drought relief, climate-change response, investment directions, institutional forms). However, to date, no comprehensive analysis with specific application to groundwater has been carried out. Some limited information with respect to the impacts of irrigation and poverty is being compiled for an FAO study (FAO, 2003b) but this will cover a mix of surface and groundwater irrigation systems.

The absence of an ability to 'manage' aquifers on a sustainable basis is no reason for inaction. Adaptive strategies that build on existing patterns of social change (e.g. where drought is used as an opportunity to move populations toward more sustainable livelihood patterns) could represent an alternative to groundwater focused 'management' strategies. In order to identify potential opportunities for this type of strategy, it is necessary to gain a better understanding of:

- who is actually affected by emerging groundwater problems (equity, etc.);

- what individuals communities and local governments are actually doing to address such problems;
- the short- or long-term coping strategies individuals follow when faced with water scarcity;
- the patterns of social and economic change that affect groundwater use and management options but that are not in themselves a response to perceived water management needs, e.g. changes in community characteristics occurring in response to globalization.

Better understanding of the above could help to develop a more realistic 'adaptive' strategy rather than a technocratic 'management' strategy for responding to groundwater problems. It should result in the identification of a wide variety of practical points for intervention.

Chapter 5
A basis for groundwater access and allocation

Introduction

Key elements to recognize in the valuation of groundwater are:

- The strategic value of groundwater (water that is located near 'high-value' uses such as urban or prime agricultural areas) as opposed to aquifers located in less strategic locations.
- Aquifers with high-quality water that is not vulnerable to pollution (an aquifer that can be reserved more readily for strategic uses may have a higher inherent 'value' than a more vulnerable one).
- The types of uses (aquifers that produce high-value environmental services, such as instream flows, may have a higher inherent value than others where such services are absent).

Access

In most situations, groundwater is a common property resource with extremely high use value. The term 'common property' refers to the status of groundwater as a resource to which all overlying landowners generally have access. The term is not intended to reflect the legal status of groundwater as a 'public', 'common' or 'private' resource.

In some societies, groundwater is or has been linked to land ownership. In others it is viewed as a 'common heritage' (not to be confused with the British 'common law' system) to which all should have equal access, at least for basic needs. These conflicting positions are enshrined in religious doctrines such as the Shari'a (where the 'right to thirst' is a basic principle) and in western legal concepts dating back to Ancient Rome (in which groundwater ownership follows land ownership). At the same time, rights of access to groundwater have generally been linked to land ownership. Thus, there is often an unclear distinction between the 'private' nature of groundwater rights and the 'public' ownership of the resource itself. This contradiction is brought to the fore by increased recognition of the need for water to be used more efficiently. Market mechanisms can play a major role in achieving efficiency objectives, and more emphasis is now being given to the nature of water as an economic resource in global debates. This emphasis translates into initiatives to clarify water rights, encourage water markets and issue 'concessions' in some countries. However, the process may be stillborn if there is no recognition that water resources, by their very 'public' nature, require regulation and are not amenable to absolute free-market solutions. As these initiatives increase, tensions related to underlying ethical issues can also be expected to rise. If groundwater is viewed as a common heritage to which all have fundamental rights of access, the ethical basis of unregulated concessions or markets that allocate water depending on ability to pay becomes controversial.

The limits of valuation methods

Estimates of the value of groundwater have assumed great importance in the context of the massive investments required to avoid pollution, remediate polluted aquifers and control overdraft. Some objective basis, related to the total economic value of the resource (including wider social and environmental values), is required to determine how much to invest in specific situations. However, the available valuation methods are partial at best.

Techniques for quantifying the economic value of groundwater resources include:

- Contingent valuation, which essentially involves asking people how much they would pay to maintain the resource or services dependent on it under carefully specified conditions.
- Hedonic pricing, e.g. obtaining a measure of the value of groundwater through differences in the value of lands with and without access to it.
- Derived demand and production cost analysis, essentially estimating the contribution of water to profits within a given set of economic activities.
- Loss analysis, estimating the value of groundwater as equivalent to the total social costs incurred when drought or depletion constrain economic activity.
- Averting behaviour in which the value of groundwater is estimated by the investments made to avoid water shortage.
- Substitution, the value of groundwater as equivalent to the least cost alternative source of supply for meeting the same set of services.

Many of the above techniques have been discussed in detail for the United States of America (National Research Council, 1997) and in the context of Jordan (Schiffler, 1998).

The above quantitative techniques do not capture adequately many of the qualitative values associated with groundwater, the full range of services and benefits that it provides. They also assume perfect knowledge of the groundwater systems in space and time and the precise impact of uses (including disposal). Non-use and *in situ* values, such as the prevention of saline intrusion and the protection of the environment against low-frequency events, are particularly difficult to establish values for. For example, valuation of ecosystem losses due to changes in groundwater conditions is an inherently subjective exercise. Some sections of society approach the environment from a spiritual and religious perspective and, in doing so, place essentially infinite value on the natural resource base that sustains environmental conditions. Other sections of society focus on specific economic activities when valuing natural resources. Weighing the relative importance to give to these different perspectives is in itself a subjective value judgement, and one that is often implicit in the valuation methodology selected. A similar ethical divide surrounds questions of ability versus willingness to pay. Quantitative measures of willingness to pay established by any of the primary methods ignore questions of ability to pay. For example, the socio-economic status of respondents can colour their responses in contingent valuation surveys. Access to groundwater is often a critical factor in the quality of life for socio-economically marginal populations in developing countries. However, their ability to pay for maintaining groundwater resources is often minimal. How this contradiction can be resolved depends on whether groundwater is treated as a common heritage (to which all have equal and fundamental rights) or as an economic good.

Beyond the ethical debate lie practical questions regarding the degree of social and technical understanding of groundwater on which value estimates are established. A basic problem concerns the interlinked and sequential nature of groundwater uses. 'Waste' from one use is often the primary source of water for a subsequent use. Tracing the chain of uses is a difficult task, as is valuing the contribution of groundwater to each one. Beyond this, it is often difficult to determine whether values reflect low-frequency but high-consequence events adequately, e.g. determining whether a value for groundwater in agriculture reflects the consequences of droughts that occur with frequencies of 20, 50 or 100 years. This relates closely to issues of information and understanding. The general public may tend to view groundwater as an underground bowl replenished regularly by rain. Values derived from this viewpoint are unlikely to reflect the finite nature of many groundwater resources or their vulnerability to pollution. Similarly, it is difficult to determine the extent to which other forms of capital can substitute for groundwater should existing uses constrain availability. The limits of substitution may only become evident when a chain of sequential and interdependent uses is understood fully. In sum, the degree to which people understand groundwater and its contributions to economic, environmental and other services may well influence the value they place on the resource.

Reconciling competition for groundwater

The aquifer systems under the most pressure from competing uses are typically those where municipal users are vying for adjacent groundwater being used by agriculture. In many instances, these systems are in arid and semi-arid zones where surface water alternatives are not available, e.g. the North China Plain and the Quetta Basin in Baluchistan, Pakistan. Reallocation is occurring, but it is generally forced, either as deeper water-supply wells capture water away from adjacent irrigation wells or as irrigation wells are closed down. The growing municipalities in the central section of the Huaihe Basin in China have been able to capture groundwater from peri-urban irrigation but not through any planned reallocation (United Nations, 1999). Reallocation of groundwater rarely occurs through the consensual transfer of use rights and in these instances it could be argued that the application of valuation methods would achieve little. Irrespective of the technical ability to capture groundwater, irrigated agriculture cannot compete with higher value municipal uses on economic grounds alone.

Concern has been voiced about the decline in water tables in key grain producing areas of the world. Brown (2002) cites China, India and the United States of America. However, others are less pessimistic. Custodio (2002) and Price (2002) call for more precise information on the hydrodynamic state of the aquifers in question.

However, this is not to say that the depletion of key aquifers in major grain producing areas is not significant. In areas of northern India and China, the resulting adjustment of local agricultural systems is causing migration from the land and social disruption (Moench, 2002). Equally, the long-term impact of aquifer drawdown is affecting surface flows and seepage, resulting in the loss of locally important wetlands. However, most of this evidence is anecdotal. Efforts to quantify the dependency of food security on groundwater using available national data can only be partial (Burke, 2002). Other more detailed attempts (FAO, in press) have illustrated the local problems with matching groundwater data and food production.

The concern that the reliance on groundwater irrigation is threatening global food security may be overplayed. Land irrigated by groundwater goes in and out of production incrementally as agricultural systems and markets respond to natural resource limits. The possibility that a shortfall in China's grain requirements would suddenly soak up the international market in traded grain may be remote, but what is troubling is the compounded effects of long-term drawdown and drought. In periods of recurrent drought when anticipated rainfall inputs are substituted by groundwater, the 'spike' in groundwater demand can result in sharp reductions at regional level. For example, stream flows in the western United States of America have been reconstructed based on dendrochronology (Meko *et al.,* 1991). Over the last 300 years, the average drought has lasted 5 years and some have lasted 15 years. Water levels in the Ogallala aquifer have been declining during periods of 'normal' or above average precipitation and the western United States is facing one of its most keenly felt periods of drought. Therefore, global food production problems could conceivably occur if there were simultaneous long-term drought across several grain producing regions where groundwater levels have already been declining. Irrespective of this, groundwater depletion, particularly in shallow aquifers, is already having direct socio-economic and environmental effects. Some of these impacts may be lagged over several years or decades as aquifer hydraulics and hydrochemistry adjust to intensive pumping regimes. However, global analysis reveals little of the inherent tensions and opportunities that are experienced with groundwater irrigation on local scales.

Chapter 6
Limits to groundwater management

Introduction

An analysis of current practices might lead to the conclusion that there is no effective system of groundwater management. It is a rare exception when wells are closed down and capped off to prevent abstraction, or limits set on pumping durations or volumes. It is not such a rare exception to observe local, consensual enforcement of pumping limits, e.g. for irrigated agriculture in Eritrea and Yemen when pumped groundwater is rationed during the dry season. These arrangements are 'customary', but have only been occasioned by the advent of inexpensive motorized pumps.

The variable patterns of groundwater use and the varied services that aquifer systems provide do not form a clear aggregate picture or status of groundwater, nor do they present an opportunity for systematic management response. Despite the highly technical work that is carried out and presented in the hydrogeological literature, the status of knowledge of the aquifer systems is often limited at the level at which a management response is required. Highly detailed studies in contaminant transport are carried out in high-value settings (usually because regulatory systems are enforced). However, accurate and reliable monitoring and regulation in the crucial aquifers of northern India (FAO, in press) and Baluchistan, Pakistan, are not available. Even if they were, it is questionable as to whether the data would provide an effective tool for regulation or furnish a clear message for the education of users and the basis for behavioural change.

Implications for integrated management

One major concern is the fact that the issues outlined above are a symptom of current water management as a whole. In general, these management practices continue to ignore the integrity of groundwater systems even in arid regions where groundwater is the 'lender of last resort'. This is particularly true in the case of large sedimentary aquifer systems that are decoupled from contemporary recharge and are effectively non-renewable. In addition, the varying scales at which groundwater systems occur and are developed or exploited pose particular management challenges beyond those of conventional surface water or river basin management. Therefore, a clear articulation of the specific guiding principles in groundwater development and criteria for evaluating policy responses to groundwater depletion and degradation is warranted. Such principles may have little to do with the more conventional principles of 'integrated water resource management', which are generally predicated on hydraulic control and regulation over river basins. This 'engineering hydrology' focus on water management continues to colour water resource management styles, which remain largely centralized and technocratic. One example is Namibia, where groundwater provides some 60 percent of bulk water, yet the institutional arrangements and investments have focused on the development of intermittent surface flows.

River basins and surface water irrigation schemes present 'neat' arrangements. The resource is naturally integrated at any point in the basin's watercourse and it is easy to monitor measurements, diversions, storage and abstractions. The impacts of irrigation abstraction and return flows occur in near real time and are immediately apparent upstream and downstream through hydraulic continuity. The system is neatly bounded, there are clear solutions of continuity, and systems of rights in use are generally clearly established. The same cannot be said for aquifer systems and groundwater development. Aquifer systems are known imperfectly, there are no clear solutions of continuity (Burke, 2000), responses are highly non-linear (geological heterogeneity and anisotropy) and can be lagged over centuries with none of the clear 'water year' rhythm observed in surface basins.

The largely technocratic, vertically integrated basin management models built around surface water schemes and the sets of incentives to surface water managers and users differ markedly from the management 'models' and incentives associated with the more imprecise character of aquifer systems and groundwater use. In addition, while the 'client base' for a basin manager would typically consist of a set number of well-identified user groups (irrigation schemes, water user associations, municipalities, etc.), the manager of an aquifer system may in practice have to engage with millions of individual users.

Therefore, the transaction costs of applying 'cross-sectoral integrated water resources management' in a classic sense can be expected to be several orders of magnitude higher. This does not appear to augur well for progressive conjunctive use management as a means to reconcile competing surface and groundwater demands. While conjunctive use is common in many irrigation schemes where individual irrigators have wells in the surface command area, the specific management of the combined sources is less common and mechanisms for doing this appear to be lacking (Kloezen, 2002; Wahaj, 2001). However, progress in applying conjunctive use and the use of extensive informal water markets in and around surface-water and groundwater irrigation schemes demonstrate the ability of irrigation end users to adapt. Such adaptation often occurs in the face of contradictory signals and incentive structures established by higher order 'managers'. Whether such de facto arrangements offer more or less equity and economic 'efficiency' is debatable. Where regulation is weak or absent, the opportunity for the richer members of a groundwater user group to capture through technology or access to land will always be there. However, it is also possible to observe enhanced equity and efficiency through the myriad of small water, energy and pumping transactions that occur among irrigation user groups (Shah, 1993).

The sometimes unclear linkages between groundwater and food production have constrained the scope for management of the resource. Given the levels of uncertainty associated with groundwater information, the broader questions that remain are whether: (i) groundwater is amenable to the same types of management approaches associated with surface water irrigation (e.g. irrigation management transfer); and (ii) it sits well within the frame of so-called 'integrated water resource management'. It is important to resolve these questions because the aquifers that are being used intensively can be expected to be in arid and semi-arid zones. Moreover, in such areas, surface water alternatives are scarce or unavailable and drawdown and pollution externalities can be expected to affect users within and outside the zone of groundwater use. Therefore, solutions that require some sort of 'integrated' or consensual and expert management (such as conjunctive use) will become imperative. However, requests to individual users to sacrifice private opportunity for the sake of basin or aquifer-wide efficiencies and equity are likely to encounter resistance, particularly where groundwater access is the principal means for accumulating assets and escaping from poverty (Moench, 2001) .

In irrigated agriculture, the scope for addressing such management problems is also conditioned by the need to serve several policy 'masters'. Irrigated agriculture is a key component of many national agriculture strategies and a factor in the political equations many national leaders face. However, it is also expected to conform to water and environmental policy initiatives. That management of groundwater use for irrigation should be part of a national and regional commitment to integrated water resource management is not in dispute, but precisely how and through which policy instruments it should be effected often remains unclear.

In terms of irrigation practice, efficiency gains are important at all scales if pressures are to be reduced on environmental flows and downstream/downgradient users. Irrigation is not in a position to foreclose on other users. It would be simplistic to assume that irrigation efficiency is not significant at basin level (Seckler, 1996). Irrespective of the impact across a particular basin or catchment system, it is the immediate deprivation of opportunity that may count. Similarly, it should also be appreciated that seeking efficiencies involves specific groups of water users and managers at the various levels. However, there may be no incentive for direct users to make efficiency gains if upstream managers cannot ensure conveyance efficiency. With groundwater, this may not apply as the incentive is generally internalized entirely by the user.

Policy response and institutional adaptation

In some cases, it is already too late to talk about the sustainable development of groundwater because the aquifers are already depleted, polluted or salinized beyond the regenerative capacity of their natural hydrogeological regimes. Some industrialized countries (e.g. the United Kingdom) are moving toward a re-examination of groundwater management in a broader political and social context (Grey *et al.*, 1995). Others (e.g. France) are maintaining a more technical perspective (Martin, 1996). Elsewhere, many developing countries that rely on pumped groundwater to sustain agricultural output and supply municipalities continue to permit the intensive levels of abstraction with little evidence of pro-active groundwater management being deployed. Technical regulation, economic incentives and participatory management approaches may offer the means to address groundwater management in the common interest. However, the character of initiatives will be determined necessarily by the local realities of the groundwater occurrences and the associated groundwater economy. Dealing with such diversity involves a different order of adaptability and flexibility than that normally associated with surface water or river basin management.

By the time groundwater arrives at the well head and enters irrigation ditches or raw-water pumping mains, it is perceived that groundwater management per se ceases and conventional water management takes over. Indeed, groundwater tends to be treated as the ultimate source of relatively high-quality water and the ultimate sink of used water. This occurs without any real appreciation of groundwater's regenerative capacity and its buffering role in the hydrological cascade. The management of surface water has fundamental implications for both groundwater quantity and quality at all stages and points of the hydrological cycle.

Therefore, it is essential to examine the scope for groundwater management not only in the strictest sense, but also as a prerequisite for integrated water resources management. This involves appealing to individual groundwater users in ways that have to do with advocacy and demonstration. It is significant that some political scientists sense a continuing tension between the 'eminence' of the State and the customary user rights of the beneficiaries (Barraque, 1998). How this tension is resolved is critical for groundwater because the management of diffuse abstraction is highly dependent upon the approach local communities take to negotiating the use of a common property. Historically, it has appeared easier, though not necessarily beneficial or cost-effective, for the State to control surface water abstractions and disposal within a vertically 'integrated' river basin plan where a central authority undertakes all the operational and regulatory functions related to the water cycle. Here the purpose of the State's intervention is the protection of the broader public interest in the basin's resources. However, this tendency in river basin management risks ignoring the important but highly distributed physico-chemical and socio-economic buffering roles of groundwater. The approach may also rely heavily on regulatory measures as opposed to economic incentives to achieve desired results. More significantly, the array of users with which basin agencies tend to engage (large sectoral user groups and local government representatives) may differ markedly from the groundwater 'stakeholders'. Millions of individual farmers are not necessarily amenable to the same degree of association that can be recognized among urban utilities, industries and command area authorities. Thus, it can be argued that integrating groundwater use within both a physical and a socio-economic framework becomes not only an environmental necessity but also a political imperative where policies of decentralization and subsidiarity are adopted.

Under these circumstances, the choice of water resource management instruments to address specific groundwater issues may still fall within the bounds defined by technical regulation, economic incentives and participation. However, they can be expected to take on a markedly different character from those associated with conventional surface water management. Equally, the quality of the information about aquifers, groundwater and user behaviour will also need to assume a specific character.

Gaps in management

Groundwater acts as the primary buffer against the impact of climate variability and spatial variability in drought. However, as human development has become more susceptible to such variability, three major gaps in groundwater management have emerged, each with significant implications for sustainable development:

- The inability to cope with the acceleration of degradation of groundwater systems by overabstraction, and effective resource depletion through quality changes (pollution, salinity).
- In general, a lack of professional and public awareness about the sustainable use of groundwater resources. In particular, a lack of coherent planning frameworks to guide all scales of groundwater development and the consequent lack of appropriate policy responses and institutional development to prevent and attenuate degradation to groundwater systems.
- The failure to resolve competition for groundwater and aquifer services between sectoral uses and environmental externalities.

These specific concerns hinge upon the central issue of awareness. This relates as much to the groundwater related environmental concerns in industrialized countries as it does to the peri-urban communities in developing countries who depend on locally available groundwater sources.

In this sense, groundwater management regimes may be expected to encompass a set of economic, regulatory and ethical levers that are operated by markets, regulators/state institutions and user associations. Effective institutional approaches need to be aware of these socio-economic realities surrounding groundwater use. They also need to appreciate the inherent risks associated with development, the level of uncertainty (plus limitations in data quality) and the range of social pressures.

Filling the gaps

In contrast to technical views of groundwater management, an overview of the Indian experience (Roy and Shah, 2002) makes the point that although indirect demand management strategies, such as rice export controls in the Punjab, do not form part of the water management debate, they may ultimately have more impact on slowing the rates of depletion than equity-based institutional solutions. This begs the question as to whether there are practical approaches for responding to groundwater problems and their socio-economic impacts that are absent in current management styles. The strategies outlined below are starting points for expanding conventional management thinking by adapting to real points of socio-economic leverage and to aquifer and user-group scales.

In general, it is possible to observe the following characteristics of current groundwater 'governance':

- Lack of data and scientific understanding limit the ability of society to predict aquifer functioning and to develop realistic rights systems.
- Rights systems are difficult to design and implement in most situations for a variety of technical and economic reasons.
- In most cases, social acceptance of private rights may be problematic.
- Aquifer management is politically complex because it would require active modification of established use patterns.
- The dynamic nature of both socio-economic globalization and global climate change makes management complex. People are increasingly mobile and often have little incentive to participate in long-term management initiatives.

From among this set of characteristics, two broad types of management approaches for groundwater emerge: (i) 'thin and wide'; and (ii) 'thick and deep'. 'Thin and wide' approaches may encompass blunt tools such as power pricing, subsidies for efficient technologies, economic policies that discourage water intensive crops, etc. They can be applied over whole countries or regions. 'Thick and deep' approaches deal with specific aquifers on the basis of command and control management whereby aquifer management targets are set and enforced through a resource regulator.

Under these circumstances, groundwater management may be most realistic when applied in a limited manner to strategic aquifers for which a social consensus supporting management exists. This can be said to have happened in the case of the Qa'Disi aquifer in southern Jordan and the sister aquifers in Saudi Arabia. However, such examples are rare. Concerning recharge initiatives in India (Shah, 2000), here a 'thin and wide' approach to resource management may ultimately prove more successful than a 'thick and deep' approach to aquifer management, such as the technocratic initiatives to aquifer recharge along the Batiah coast in Oman and in the Quetta Basin in Baluchistan, Pakistan. In any event, the impacts of groundwater management approaches require specific monitoring and evaluation periods in order to make an accurate assessment of success in terms of aquifer response alone. It is doubtful whether enough time has elapsed to allow such an assessment of the various initiatives and to identify clear bright spots. These considerations apart, focusing management on strategic aquifers would also allow society to concentrate the required scientific, monitoring and enforcement tools on relatively small areas. In addition, any current users displaced by management could be absorbed more easily than if management were attempted over larger areas. The design of implementation strategies to initiate and then propagate across the area of concern will be key. The recharge movement observed by Shah (2000) will deserve attention where it is clear that such viability may have more to do with social structures than the technical feasibility of conservation and regulation.

In order to start addressing gaps in management, it is important to recognize that institutional innovation and adaptation will need to be more sensitive to the range of influences and management instruments. A diagnostic to develop such adaptations will need to cover:

- macroeconomic policies;
- sector policies;
- rights systems;
- institutions and capacities;
- regulatory frameworks;
- public involvement.

Against the 'soft' institutional strategies, it is possible to define sets of technical options that relate directly to groundwater. Although these options may present expanded opportunities to manage groundwater, they would have to be applied strategically in circumstances that are amenable (where uptake of technical strategies will succeed). Such technical options include:

- conjunctive management (conjunctive use and aquifer storage and recovery);
- conservation enhancement and protection;
- water harvesting and supply enhancement;
- irrigation efficiency improvement and demand management.

The design of a suite of institutional and technical strategies and their implementation at the scale required to make an impact (to conserve or reallocate groundwater resources) is unlikely to be achieved in the short term. Countries reliant upon groundwater irrigation, such as Mexico, are now considering periods of decades (H. Garduno, personal communication). It is up to agriculture to adapt and improve its water productivity and so release the pressure before the resource base is exhausted. In part the agriculture sector can achieve this through improvements in crop yields, irrigation efficiency and post-harvest processing. However, it may be non-technical interventions that play a greater role. Resource substitution through food imports and transfers to higher value uses are already making their mark. Setting up the socio-economic tools for more flexible allocation of scarce resources will be pivotal to addressing equity and efficiency concerns. For example, improved systems of land tenure and water use rights can help considerably in spreading local water risk and improving food security by optimizing local food production patterns. Equally, the economic signals to food producers need to be clear and stable. Indeed, sound agriculture policy can set a strategic balance between rainfed and irrigated agriculture where there is effective demand and real comparative advantage in domestic resource costs. Furthermore, agriculture policies that encourage different forms of agriculture in different areas could be critical. For example, India has nationwide policies governing the support price for rice and wheat. Such national-level policies could be tailored to reflect regional water availability differences. Thus, they could encourage production of water intensive and less water intensive crops in areas with different comparative advantages. Here again, 'thin and wide' approaches can complement 'thick and deep' approaches.

This is where there is real scope for savings in agricultural water demand. However, it will involve a change in the balance between rainfed and irrigated agriculture and a number of related socio-economic changes. People are making social transitions as a result of many other factors that are not related to water resource availability or management. Where people are making such a transition and the groundwater resource base is not being used in a sustainable manner, policies can be tailored in ways that encourage transitions towards low water-use livelihoods (often non-agricultural). This may not be part of conventional approaches to groundwater management, but it could be key where direct enforcement of water management targets is deemed impossible in either technical or political terms.

Chapter 7
Conclusions and recommendations

CONCLUSIONS

This paper highlights three key points. First, access to groundwater will continue to allow intensification of agricultural production in response changing patterns of demand. Second, the scope for managing agricultural demand for groundwater is limited, particularly where rural communities are trying to escape from poverty. Third, the overexploitation of aquifers by agriculture is forcing users into economic and social transitions (moving off the land or transferring resources or user rights to competing users, e.g. municipal and industrial users). The net consequences will be:

- A loss of some strategic aquifers.
- An enhancement of agricultural productivity in relation to overall water use and uptake of conjunctive use.
- A marked transfer of groundwater from agriculture to other competing users.
- Substitution of groundwater by imports or alternative sources.

The expansion of irrigated agriculture in the twentieth century has decoupled the water user from the inherent risk of exploiting annually renewable water resources. The apparent reliability of storage and conveyance infrastructure and the relative cheapness and flexibility of groundwater exploitation offered by mechanized drilling and pumping have allowed groundwater irrigation to take up opportunities in the continuum between rainfed and full control irrigation. This has sheltered the end user and society as a whole from the risk of crop failure due to natural hydrological variability. The imperative for in-field irrigation efficiency has been partially removed. This is because the physical and economic management of the resource is often determined by command area authorities or, in the case of groundwater pumping, by the performance of power utilities, which have no direct interest in water resource conservation. As a result, the resource base has been degraded, and irreparable damage has occurred in some cases. It is argued that the rigidity of water resource management in many irrigation systems is not attuned to the inherent variability of the natural systems upon which they depend. Furthermore, irrigation management systems can work in a more balanced way by spreading risk equitably and transparently amongst the resource regulators, managers and users. However, this requires a more flexible approach to natural resource management, one that is conditioned not only by natural parameters, but also by the socio-economic settings.

Groundwater will continue to be used intensively and some expansion of irrigated agriculture can be expected to develop new groundwater sources, particularly as markets for agricultural produce change. This will happen in parallel with land going out of irrigated production as a consequence of physical depletion, migration of low-quality water, economic depletion (where pumping costs become excessive), waterlogging and salinization, and groundwater transfers out of agriculture (e.g. in the western United States of America).

These changes will be incremental. Hence, the scenario proposed by Brown (1998) is unlikely to occur. However, there is potential for widespread drought to occur in conjunction with groundwater declines. If the impacts of intensive use are incremental, so too is recovery, but only if systems can be relaxed. However, with over 100 years of development in the Ogallala aquifer, a collective agreement to comanage a common property aquifer only managed to attenuate the rate of decline, not reverse it (White and Kromm, 1995). In this sense, effecting agricultural and social transition, of which water management is a part, may provide more scope for relieving pressure on key aquifers rather than relying on water management alone.

Recommendations

Collaborative initiatives in groundwater management

The results of this overview suggest two broad avenues for future work. The first avenue involves the development of a detailed research programme to gather groundwater data directly from governments and other sources within key countries in order to develop an improved picture of groundwater use and conditions. This type of picture is a prerequisite for developing: (i) more informed assessments of the implications of groundwater conditions for food security; and (ii) scientifically founded courses of action for managing the resource base. The second avenue focuses on the development of adaptive responses to water problems and policy approaches that reflect and respond to uncertainty, change and the absence of real understanding of systems and their interactions. Inherent limitations in the nature of scientific information in conjunction with the dynamic process of social and institutional change occurring in many parts of the world make this second avenue at least as important as the first.

In addition to these two broad areas, the analysis suggests a variety of key points of leverage for technical assistance organizations to assist in developing effective responses to emerging groundwater problems. These points of leverage are listed below in a particular sequence for specific reasons. Effective responses to emerging groundwater problems are essential. However, this paper argues that the viability of traditional integrated management approaches is limited by a wide array of data, technical, social and political factors. As a result, society needs to proceed on two equally important courses.

An interagency working group on groundwater management would allow the UN agencies concerned to focus efforts along these two courses and develop mutually reinforcing normative and operational programmes.

FAO, the IAEA, UNESCO and the UNDESA have contributed to this initiative, but the group needs to include agencies with a declared comparative advantage in groundwater science, information and management. The WHO can provide specific direction on groundwater pathogens/remediation, the World Bank on investment in groundwater and the Regional Commissions on specific regional perspectives.

Such an interagency group could be responsible for collating the UN-agency thinking (including that of their official partners in the Consultative Group on International Agricultural Research and with international NGOs) and contributions in the areas outlined below.

Rethinking the approach to groundwater management

The need to rethink the approach to groundwater management is a theme that runs throughout much of this paper. Standard management approaches depend heavily on the presence of basic data and on institutional capacities for regulation, scientific research, etc, that are absent in many countries. Because such capacities and data often require decades to develop, alternative approaches are essential in order to address the types of problems now emerging in many regions. Furthermore, this research suggests that strategies that build off existing trends within society or help populations to adapt may be as effective as strategies that attempt to manage the groundwater resource base directly. Further research to clarify existing coping mechanisms and to identify or test the viability of adaptive strategies could represent a major starting point for an initiative to 'rethink groundwater'. The development of criteria suggesting where traditional forms of groundwater management may or may not be possible is also a key area for work. This could be of critical importance to governments and other actors seeking to identify locations where different approaches are likely to prove viable.

Basic research

Basic research on groundwater is fundamental to any attempt to manage groundwater or respond to the problems now emerging in many areas. For example, in India, seasonal water-level fluctuations (whether natural or related to extraction) may have implications for water access that are at least as important as the presence or absence of 'overdraft' conditions. Large-magnitude fluctuations may be particularly important in hard-rock regions where storage is low. There has been relatively little research on groundwater availability in hard-rock regions because they typically form poor aquifers. However, they do represent a major source of water for farmers in locations such as India. Therefore, better understanding of groundwater dynamics in hard-rock areas is important.

In addition to specific environments, it is important to conduct further research to identify techniques for the rapid and accurate evaluation of water-balance components under developing-country conditions. In many regions, even order-of-magnitude estimates for extraction, recharge, evapotranspiration, etc. are unavailable. Research that would help to 'tighten' and improve water-balance models with the types of data available in developing countries is important. Ultimately, such research may be able to move models away from the current uncertainties.

Beyond groundwater per se, further research focused on the changing social context in with overdraft problems are emerging appears important. Understanding the implications of groundwater overdraft for food security and livelihood sustainability requires detailed understanding of the way in which rural agricultural societies are evolving and of the coping strategies rural populations have developed to deal with water scarcity. Whether or not people are actually able to shift livelihoods away from agriculture into other equally or more productive strategies is of fundamental importance to understanding the impact that overdraft may have on them. This type of research is essential to determine whether or not 'adaptive' strategies can meet the dual objectives of improving livelihoods while increasing the sustainability of basic groundwater resources. It is also essential in order to identify points of leverage where governments or other organizations could assist rural populations to adapt to emerging water problems.

Groundwater monitoring and data collection

Many governments and other actors attach a low priority to collecting basic data on resource conditions. However, such long-term data covering all key elements of the hydrological cycle including groundwater fluctuations and water-level trends are essential as a basis for management and for evaluating the implications of changes in use. In addition, models based on such data can play a central role as 'negotiating texts' where conflicts over resources or their management emerge within society. In the absence of such data, the parties have little basis for reaching agreement on the actual nature of groundwater systems. As a result, debates over management have little hope of reaching closure. Because data provide the foundation for social agreements regarding how aquifer systems work or the actual amount of water available, they can serve as a key tool of conflict resolution. Therefore, continued support for basic data collection and groundwater evaluation is justified on both scientific and social process grounds.

Data dissemination and access

Data access is probably the single most important factor determining the ability of social auditors (e.g. NGOs and other civil society actors) to press governments and society as a whole to address emerging problems and their social or environmental impacts. As a result, activities that support data dissemination remain a key point for action. However, while comprehensive initiatives such as the FAO Aquastat database (accessible at: (http://www.fao.org/ag/AGL/aglw/Aquastatweb/Main/html/aquastat.htm) provide a national breakdown of groundwater-dependent irrigation, attempts to refine this breakdown will encounter the data problems demonstrated in this report. Therefore, continued support for the dissemination of national groundwater data for groundwater users, where available, would seem a more appropriate direction to take.

Integrated management in strategic locations

Together with the need to rethink groundwater and identify new strategies to address emerging problems, continued efforts to implement integrated groundwater management using more standard regulatory and economic approaches are equally important in locations where such approaches appear viable. Because standard management approaches tend to require substantial technical support and often involve politically or economically difficult decisions, success may depend on focusing management initiatives in areas of particular strategic importance. For example, aquifers that serve as the primary source of freshwater supply for urban areas or support critical environmental values may represent strategic locations on which to focus management efforts. In most countries, such aquifers represent a small fraction of total groundwater use. They are also likely to involve uses where it is relatively easy to generate broad consensus within society regarding the importance of management and aquifer protection. As a result, approaches that focus management on such strategic locations are more likely to be successful than efforts to manage groundwater throughout broad regions.

Laying the foundations for management in complex locations

Adaptive strategies are of equal importance and complementary to more standard groundwater management approaches. In many cases, they can provide the breathing space necessary to develop the institutions and information essential for more focused management. Thus, it is important to continue to lay the foundations for direct groundwater management even where it may not produce results in the short to intermediate term. Therefore, continued FAO support for basic groundwater data collection, the development of legal frameworks to enable management and the development of supporting organizations is important.

Disseminating global lessons

A final key point of leverage for UN-system agencies lies in the global perspective they can bring to groundwater based on actual national data sets. Governments and communities in many parts of the world are trying different approaches to groundwater monitoring, analysis and management. Harvesting and disseminating the lessons from these initiatives could serve as a catalyst for the development of approaches that are effective even in the most difficult locations. As a result, activities that support the harvesting and dissemination of instances of adaptive groundwater management (simply to show what happens) will continue to be an important activity for UN agencies involved in groundwater management. The actual experience of groundwater management, or the lack of it, needs to be charted if real responses are to be effective.

Target studies could include case studies of:

- groundwater management failure;
- specific socio-economic impacts resulting from overabstraction and pollution;
- intense competition among water users (private and public);
- intersectoral competition between irrigated agriculture and urban water supplies;
- competition between communities located at recharge and discharge areas of aquifer systems;
- competition over transboundary aquifers (exploitation and pollution).

Support for the World Water Assessment Programme

The main objective of the World Water Assessment Programme (WWAP) is to develop the tools and skills needed to achieve a better understanding of the basic processes, management practices and policies that will help improve the supply and quality of global freshwater resources. Thematic advocacy and guideline documents concerning rethinking and improving groundwater management, planned in the framework of a joint UN Agency Programme, should support the WWAP. In particular, its objective "identifies water management strategies and policies which work well and those which are unsatisfactory and analyses the reasons for success and failure" is closely related to planned activities of UN Agency Programmes. The Interagency Working Group could prepare a timetable and working plan for cooperation on the World Water Development Report, which is one of primary activities of the WWAP. The development of the methodological framework of groundwater resources indicators and indices to monitor water-related social and environmental performance (and their testing on target case studies) is a time-consuming process. The cooperation of the Interagency Working Group will be very useful. At the ACC/SWR meeting in Tokyo (April 2001), UNESCO was given the coordinating role for the development of groundwater resources indicators.

References

Barraque, B. 1998. Groundwater management in Europe: regulatory, organisational and institutional change. *In* T. Hilding-Rydevik & I. Johansson, eds. *How to cope with degrading groundwater quality in Europe*. Stockholm.

BGS. 1995. Overexploitation of Aquifers - Final Report. British Geological Survey Technical Report WC/95/3. Keyworth, Nottingham, UK.

Brown, L. & Halweil, B. 1998. China's water shortage could shake world food security. *Wld. Watch Mag.*

Brown, L. 2002. *Rising temperatures and falling water tables raising food prices record world grain harvest shortfall of 83 million tons in prospect.* Earth Policy Institute Web site, 21 August 2002 (available at http://www.earth-policy.org/Updates).

Burke, J.J. & Moench, M. 2000. *Groundwater and society, resources, tensions and opportunities. themes in groundwater management for the 21st century.* New York, United Nations. 170 pp.

Burke, J.J. 1996. Hydrogeological provinces in central Sudan: morphostructural and hydrogeomorphological controls. *In* T. Brown, ed. *Groundwater and geomorphology*. Chichester, UK, Wiley.

Burke, J.J. 2000. Land and water systems: managing the hydrologic risk. *Nat. Res. For.,* 24: 123-136.

Burke, J.J. 2002. Groundwater for irrigation: productivity gains and the need to manage hydro-environmental risk. *In* R. Llamas & E. Custodio, eds. *Intensive use of groundwater challenges and opportunities*. 478 pp. Abingdon, U.K., Balkema.

Central Ground Water Board (CGWB). 1995. *Groundwater resources of India*. Faridabad, India, Central Groundwater Board, Ministry of Water Resources, Government of India.

Chaturvedi, M.C. 1976. *Second India studies: water*. New Delhi, Macmillan Company of India Limited.

Custodio, E. 2002. Aquifer overexploitation – what does it mean? *Hydro. J.*, 10 (2): 254-277.

Edmunds, W.M. & Milne, C.J. 2001. *Paleowaters in coastal Europe: evolution of groundwater since the late Pleistocene*. Geological Society Special Publication No. 189. London. 332 pp.

Edmunds W.M. and Wright, E.P 1979 Groundwater recharge and paleoclimate in the Sirte and Kufra Basins, Libya. J. Hydrol (40)

Edmunds, W.M., Darling, W.G. and Kinniburgh, D.G. 1987. Estimation of aquifer recharge using geochemical techniques. Final Report of the Lower Atbara River Basin Project. BGS/NERC Report WD/OS/87/1

FAO. 1997. *Irrigation in the Near East Region in figures.* Waer Reports No. 9. FAO, Rome.

FAO. 2003a. *Re-thinking the Approach to Groundwater and Food Security.* Water Reports No. 24. Rome.

FAO. 2003b. Preliminary review of the impact of irrigation and povery with special emphasis on Asia. AGL/MISC/34/2003. Rome.

Grey, D.R.C. *et al.* 1995. *Groundwater in the UK. A strategic study. Issues and research needs.* Groundwater Forum Report FR/GF 1. Marlow, UK, Foundation for Water Research.

Ilich, N. 1996. *Ganges River Basin Water Allocation Modelling Study*. Calgary. Consultant Report to the World Bank.

Joshi, P.K., Tyagi, N.K. & Svendsen, M. 1995. Measuring crop damage due to soil salinity. *In* M.S.A. Gulati, ed. *Strategic change in Indian irrigation*. New Delhi, Macmillan India Limited.

Kahnert, F. & Levine, G. 1989. *Key findings, recommendations, and summary. Groundwater irrigation and the rural poor: options for development in the Gangetic basin.* Washington, DC, The World Bank.

Kloezen, W.H. 2002. *Accounting for water. Institutional viability and impacts of market-oriented irrigation interventions in Central Mexico*. Wageningen University, The Netherlands. 291 pp.

Lerner, D. et al. 1990. *Groundwater Recharge. A Guide to Understanding and Estimating Natural Recharge*. International Association of Hydrogeologists. International Contributions to Hydrogeology Volume 8Hannover, Heise 345pp..

Martin, Y. 1996. Rapport sur la gestion durable des eaux souterraines. Paris, Conseil Général des Mines, Ministère de l'Industrie, de la Poste et des Télécommunications.

Meko, D.M., Hughes, M.K. & Stockton, C.W. 1991. Climate change and climate variability: the paleo record. *In*: *Managing water resources in the west under conditions of climate uncertainty*, pp. 71-100. Washington, DC, National Academy Press.

Ministry of Water Resources. 2001. *China agenda for water sector strategy for north China*. Volume 2: Main Report April 2, 2001 Report No. 22040-CHA. 309 pp.

Moench, M. 2002. Groundwater and Poverty: Exploring the Connections. *In* R. Llamas & E. Custodio, eds. *Intensive use of groundwater challenges and opportunities*. 478 pp. Abingdon, U.K., Balkema.

Moench, M. 1996. *Groundwater policy: issues and alternatives in India*. Colombo, International Irrigation Management Institute.

Moench, M. and H. Matzger. 1994. Ground Water Availability for Drinking in Gujarat: Quantity, Quality and Health Dimensions. *Economic and Political Weekly* XXIX (13): A-31-A-14.

Morris, B. L., *et al*. 1994. *The Impact of Urbanisation on Groundwater Quality* (Project Summary Report). Keyworth, British Geological Survey.

National Research Council. 1997. Valuing ground water. Economic concepts and approaches. Washington, DC, National Academy Press. 189 pp.

Pedley, S. and Howard G. 1997. The public health implications of microbiological contamination of groundwater. *Quarterly Journal of Engineering Geology* 30 (2): 179-188.

Phadtare, P. N. 1988. Geohydrology of Gujarat State. Ahmedabad, Central Groundwater Board, West Central Region: pp 103 + Appendices.

Postel, S. 1999. *Pillar of sand: can the irrigation miracle last?* New York, USA, Norton.

Price. M. 2002. *Who Needs Sustainability?* In Hiscock, K.M. et al. eds. Sustainable Groundwater Development. Geological Society Special Publication No. 193. London.

Repetto, R. 1994. *The "second India" revisited: population, poverty, and environmental stress over two decades*. Washington, DC, World Resources Institute.

Rogers, P., P. Lydon, et al. 1989. *Eastern Waters Study: Strategies to Manage Flood and Drought in the Ganges-Brahmaputra Basin*. Arlington, Irrigation Support Project for Asia and the Near East, USAID Project No. 3-7631510.

Roy, A. D. and T. Shah. 2002. Socio-ecology of groundwater irrigation in India. *Intensive Use of Groundwater: Challenges and Opportunities*. R. Llamas and E. Custodio. Lisse, A.A. Balkema: 307-335

Sadeque, S. Z. 1996. *Nature's Bounty or Scarce Commodity -- Competition and Consensus Over Ground Water Use in Rural Bangladesh.* Annual Conference of the International Association for the Study of Common Property, University of California, Berkeley.

Sampal, P. 2000. *Deep Trouble: The Hidden Threat of Groundwater Pollution.* WorldWatch Paper 154. 55pp.

SASS/OSS. 2001. *Acquisition, organisation and analysis of the SASS hydrological data.* Tunis, OSS.

Schiffler, M. 1998. *The economics of groundwater management in arid countries.* GDI Book Series II. London, Frank Cass. 394 pp.

Seckler, D. 1996. *The new era of water resources management.* Research Report 1. Colombo, IIMI.

Shah, T. 1993. *Groundwater markets and irrigation development: political economy and practical policy.* Mumbai, India, Oxford University Press.

Shah, T. 2000. *Mobilizing social energy against environmental challenge: understanding the groundwater recharge movement in western India.* Natural Resources Forum. 24 (1).

Shah, T. 2001. *Wells and welfare in the Ganga basin: public policy and private initiative in eastern Uttar Pradesh, India.* Research Report 54. Colombo, IIMI.

Simmers, I. ***et al.*** **Eds**. 1992 Selected papers on aquifer overexploitation. Puerto de laCruz. Tenerife, Spain April 15-19 1991. International Association of Hydrogeologists, Heise.

United Nations. 1960. *Large scale groundwater development.* New York, Water Resources Development Centre.

United Nations. 1999. *Perspectives on integrated basin management for a policy study on the Huaihe river basin, China.* Report of the Department of Economic and Social Affairs.

United Nations. 2000. *Groundwater and society: resources, tensions and opportunities.* New York, United Nations Department of Social and Economic Affairs, Institute for Social and Environmental Transition. 170 pp.

Vaidyanathan, A. 1994. *Second India studies revisited: food, agriculture and water.* Madras, India, Madras Institute of Development Studies.

Wahaj, R. 2001. *Farmer actions and improvement in irrigation performance below the Mogha. How farmers manage water scarcity and abundance in a large scale irrigation systems in south-eastern Punjab, Pakistan.* Wageningen University, The Netherlands. 245 pp.

White, S.E. & Kromm, D.E. 1995. Local groundwater management effectiveness in the Colorado and Kansas Ogallala Region. *Nat. Res. J.*, Vol. 35: 275-307.

World Bank. 1998. *Groundwater in Urban Development: Assessing Management Needs and Formulating Policy Strategies.* World Bank Technical Paper No. 390. Washington, DC.

World Bank. 1999. *Groundwater in rural development: facing the challenges of supply and resource sustainability.* World Bank Technical Paper No. 463. Washington, DC.

WRAY-35. 1995. *The water resources of Yemen.* Sana'a, General Department of Hydrology; The Netherlands, TNO Institute of Applied Geoscience.

FAO TECHNICAL PAPERS

WATER REPORTS

1	Prevention of water pollution by agriculture and related activities, 1993 (E S)
2	Irrigation water delivery models, 1994 (E)
3	Water harvesting for improved agricultural production, 1994 (E)
4	Use of remote sensing techniques in irrigation and drainage, 1995 (E)
5	Irrigation management transfer, 1995 (E)
6	Methodology for water policy review and reform, 1995 (E)
7	Irrigation in Africa in figures/L'irrigation en Afrique en chiffres, 1995 (E/F)
8	Irrigation scheduling: from theory to practice, 1996 (E)
9	Irrigation in the Near East Region in figures, 1997 (E)
10	Quality control of wastewater for irrigated crop production, 1997 (E)
11	Seawater intrusion in coastal aquifers – Guidelines for study, monitoring and control, 1997 (E)
12	Modernization of irrigation schemes: past experiences and future options, 1997 (E)
13	Management of agricultural drainage water quality, 1997 (E)
14	Irrigation technology transfer in support of food security, 1997 (E)
15	Irrigation in the countries of the former Soviet Union in figures, 1997 (E) (also published as RAP Publication 1997/22)
16	Télédétection et ressources en eau/Remote sensing and water resources, 1997 (F/E)
17	Institutional and technical options in the development and management of small-scale irrigation, 1998 (E)
18	Irrigation in Asia in figures, 1999 (E)
19	Modern water control and management practices in irrigation – Impact on performance, 1999 (E)
20	El riego en América Latina y el Caribe en cifras/Irrigation in Latin America and the Caribbean in figures, 2000 (S/E)
21	Water quality management and control of water pollution, 2000 (E)
22	Deficit irrigation practices, 2002 (E)
23	Review of world water resources by country, 2003 (E)
24	Rethinking the approach to groundwater and food security, 2003 (E)
25	Groundwater management: the search for practical approaches, 2003 (E)

Availability: February 2003

Ar	– Arabic	Multil	– Multilingual
C	– Chinese	*	Out of print
E	– English	**	In preparation
F	– French		
P	– Portuguese		
S	– Spanish		